Nitrogen metabolism in pla

Nitrogen metabolism in plants

C. M. BRAY

Longman
London and New York

Longman Group Limited
Longman House Burnt Mill, Harlow,
Essex CM20 2JE, England
Associated companies throughout the world

Published in the United States of America
by Longman Inc., New York

© Longman Group Limited 1983

First published 1983

British Library, Cataloguing in Publication Data
Bray, C.M.
   Nitrogen metabolism in plants.
   1. Plant cells and tissues
   2. Nitrogen metabolism
   I. Title
   581.87'61   QK725

ISBN 0–582–44640–6

Library of Congress Cataloguing in Publication Data
Bray, C.M., 1945-
   Nitrogen metabolism in plants.

   Bibliography: p.
   Includes index.
   1. Nitrogen—Metabolism. 2. Plants—Metabolism.
I. Title.
QK898.N6B7 1983   581.1'335   82–8942
ISBN 0–582–44640–6   AACR2

Set in Linotron 202 Times.
Printed in Singapore by
Four Strong Printing Co.

Contents

Contents

Preface

The widespread distribution of nitrogen in a multitude of different molecules throughout the plant kingdom is reflected in the vast amount of literature available on specialised aspects of nitrogenous compounds and their metabolism; but often the student does not have the depth of knowledge of the subject to be able to assimilate the exciting information contained in these advanced treatises. The aim of this text is to furnish the reader with a basic grounding in the metabolism of nitrogen-containing compounds in plant cells including the control of these processes, and to encourage the reader to further his or her interests in each of the areas discussed via the recommended 'suggestions for further reading' found at the end of each chapter.

The overriding aim of the text has been to present a comprehensive survey of the topic in as clear and concise a manner as possible, demonstrating the metabolic interrelationships between nitrogen-containing compounds in plants and the various ways in which the plant cell controls these numerous metabolic processes. Wherever possible an attempt has been made to summarise the metabolic pathways discussed in the text in a clear, diagrammatic form. The book begins by examining the fixation of atmospheric dinitrogen into organic linkage and proceeds through the biosynthesis of 'small' organic nitrogen-containing compounds (amino acids, purines, pyrimidines, etc.) through to the synthesis of the cell's genetic material, the control of gene expression, ending with a survey of nitrogen turnover in the whole plant during the various stages of seed development, germination, plant growth and senescence. Although the book is arbitrarily divided into chapters it is hoped that this will not 'compartmentalise' the various aspects discussed in these chapters in the mind of the reader. Indeed, the objective of the text is to convey the unity and interdependence of the metabolic processes discussed at both the cellular level and at the level of the whole plant. With this view in mind, the main function of the final chapter is to integrate all the basic pathways, controls and concepts discussed in the earlier chapters so that their interdependence, operation and control in an integrated fashion in the whole plant during various developmental stages can be seen clearly.

It is my hope that this text will provide the student with an insight into how much, or indeed how little, is known of the various processes involving the metabolism of nitrogen in plants and furthermore stimulate the desire to examine in greater depth those areas of plant biochemistry which are fundamental to progress in both the pure and applied aspects of research in our rapidly changing world.

Manchester 1982   CMB

Acknowledgements

I should like to thank Professor G.R. Barker for his encouragement when I began writing this text. I am also happy to acknowledge the many helpful discussions held with other members of the Biochemistry Department in the University of Manchester, especially Sheila Standard and Dale Smith who have kept the momentum of research going during the period in which this book was written. I would like to thank Professor Dennis Baker for his useful criticisms and helpful suggestions and the staff at Longman for their help and interest. A debt of gratitude is also owed to Miss Mandy Taylor for her excellent preparation of the typescript. Finally, it would indeed be unfair if I did not record my appreciation of a most understanding son, Michael, who has tolerated, with great patience, his father's absence from more familiar pursuits during the past several months.

1

Interconversions of nitrogen

Plants growing on land exist in a gaseous atmosphere which is 80 per cent dinitrogen (N_2). Given this relative abundance of the element nitrogen it is surprising to find that in the agricultural areas of our planet, where sunlight and water supply are not limiting for crop growth, productivity is determined mainly by the availability of forms of nitrogen in the soil which can be utilised by the plant for its growth process. This apparent anomaly is explained by the fact that the dinitrogen of the atmosphere is chemically very unreactive and a crucial step in the utilisation of nitrogen by plants involves its conversion into more reactive forms, usually the highly oxidised states (nitrate or nitrite) or the reduced state (ammonia). This *fixation* of nitrogen is performed by those micro-organisms which possess the enzymes required to reduce the nitrogen atom to ammonia under the relatively mild conditions which exist in the living cell. In these inorganic forms the nitrogen atom can be absorbed by the roots of the plant and subsequently be assimilated into the many important organic nitrogen-containing compounds, such as amino acids, proteins and nucleic acids, found in plants. It is unlikely that organic nitrogenous compounds found in the soil represent a major nutritional source of nitrogen for higher plants and the plant relies on ammonia, or more often nitrate in the soil as its prime nitrogen source for nutrition.

The nitrogen cycle

If this unidirectional conversion of inorganic nitrogen into organic nitrogen by plants were to continue indefinitely then the atmosphere and soil would soon become depleted of inorganic forms of nitrogen. However, in nature there exists a series of interrelated processes termed the nitrogen cycle (Fig. 1.1) which describes the recycling of the nitrogen atom through its various inorganic and organic forms in the biosphere.

Atmospheric nitrogen can be 'fixed' into ammonia and then taken up by the roots of the plant and *assimilated* into organic nitrogen-containing compounds intracellularly. Alternatively it can be converted into nitrate by the nitrifying bacteria of the soil in a process known as *nitrification* and taken up by the plants as nitrate, the form preferred by plants, prior to reduction to ammonia within the plant cell and subsequent assimilation into

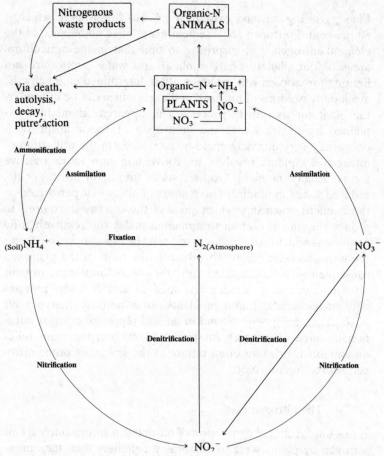

Fig. 1.1 The biological nitrogen cycle

organic nitrogen-containing compounds. Several types of denitrifying bacteria can also reduce the nitrate or nitrite of the soil to dinitrogen by utilising these inorganic ions as a substitute for oxygen in respiration. This *denitrification* leads to a net loss of utilisable fixed nitrogen from the soil.

The organically bound nitrogen found in plants and subsequently animals which utilise plants as a food source remains largely in the reduced organic form in the cell and is returned to the cycle via the process of *ammonification* which occurs on death or decay of these organisms and of their nitrogenous waste products, resulting in the production of inorganic nitrogen in the form of ammonia which can then re-enter the cycle. Ammonification involves hydrolysis of the polymeric organic nitrogen-containing compounds found in living cells into their monomeric components which can then be utilised in microbial respiration or fermentation resulting in the liberation of ammonia. In the absence of any external factors such as the application of nitrogenous fertiliser the processes of nitrogen fixation, microbial decomposition, nitrification and denitrification will determine the soil nitrogen condition.

Nitrogen fixation

In the fixation of nitrogen by biological systems the dinitrogen of the atmosphere is converted initially into ammonia, which in most instances is rapidly assimilated by the metabolic processes of the cell into organic nitrogen-containing compounds. It has been estimated that between 100 and 200 million tonnes of nitrogen are fixed in this way each year, a figure which represents more than 70 per cent of the input into the world's soil and water nitrogen. Of the remaining 25–30 per cent, nitrogenous fertiliser contributes approximately 15 per cent of the input and lightning, u.v. irradiation, the internal combustion engine and electrical sparking phenomena, which produce oxides of nitrogen, contribute the remaining 10 per cent of the nitrogen input. Through the fixation of nitrogen by biological and non-biological systems, it is possible to compensate for the net loss of fixed nitrogen to the atmosphere incurred via the denitrification steps of the nitrogen cycle. It is important to realise that none of the dinitrogen of the atmosphere or upper layers of the earth's surface is utilisable by plants until it has been fixed into ammonia and subsequently nitrate, etc.

The organisms which possess the capacity to fix nitrogen are invariably the most primitive living organisms, namely bacteria (including blue-green bacteria), yet even in these organisms, the nitrogen-fixing ability is by no means universally distributed. It is possible to divide nitrogen-fixing systems into two categories:

1 fixation performed by free-living organisms;
2 fixation performed by symbiotic associations between free-living systems and plants.

Of the free-living systems, bacteria are dominant among nitrogen-fixing organisms of the soil while blue-green bacteria appear to have the chief nitrogen-fixing role in aquatic environments. Symbiotic systems exist between bacteria (in particular, species of *Rhizobium*) and legumes, micro-organisms and non-leguminous plants, algal symbioses, and a group of systems collectively termed associative symbioses in which there is some interdependence between the bacteria and the plant (principally grasses) although both can grow satisfactorily apart.

Free-living systems

Bacteria belonging to many different genera which grow under markedly different conditions possess the ability to fix nitrogen (Table 1.1), but these bacteria do not normally fix nitrogen if combined (fixed) nitrogen is present in the environment. Also when fixing nitrogen, ammonia – the primary product of nitrogen fixation – is nearly always assimilated as fast as it is formed and only on death and decomposition of the microbe, when ammonification has taken place, does the fixed nitrogen become available to plant systems. It is difficult to evaluate the amount of nitrogen fixed by free-living bacteria, but it has been estimated that these organisms are 1,000 times less effective in contributing

Table 1.1   Examples of nitrogen-fixing bacteria

	Examples of genus or type	Examples of species
Aerobes	*Azotobacter*	*A. chroococcum, A. vinelandii*
	Azotococcus	*A. agilis*
Strict anaerobes	*Clostridium*	*C. pasteurianum*
Facultative (only fix N_2 when growing anaerobically	*Klebsiella*	*K. pneumoniae*
	Bacillus	*B. polymyxa*

fixed nitrogen to the soil than a good symbiotic leguminous association.

From an ecological viewpoint, blue-green algae are the most important phototrophs, making use of light energy to fix carbon dioxide. These organisms utilise solar energy for the nitrogen fixation process so that, providing they are not in the dark, energy ceases to be a critical limitation. This is not the case for the ordinary bacteria which possess the capability for nitrogen fixation since these organisms rarely receive sufficient energy-providing substrates in nature to enable them to upgrade the nitrogen content of the environment by 'exporting' fixed nitrogen produced in excess of normal cellular requirements. Most of the free-living blue-green algae fix nitrogen in specialised cells called hetero-cysts, which serve to exclude oxygen, a potent inactivator of the proteins involved in the nitrogen fixation process, from the site of nitrogen fixation. Those members of this group which do not produce heterocysts will only fix nitrogen when there is a low partial pressure of oxygen in the medium and a low level of illumination. The blue-green bacteria are usually the first organisms to colonise arid and infertile regions and although their major ecological contribution to the nitrogen cycle is made symbiotically they still contribute, as free-living organisms, several hundred-fold more fixed nitrogen to the soil than do free-living bacteria, such as *Clostridium* and *Azotobacter*.

Symbiotic systems – associations of micro-organisms with plants

From the agricultural viewpoint, the most important nitrogen-fixing systems are those which are undertaken in association with a plant. The reasons for this are two-fold:

1 nitrogen is fixed close to the plant roots where it is needed;
2 even though some of the ammonia produced as a result of fixation may be used by the microbial symbiont to support its own cell maintenance and growth, a large proportion is exported into the host cell, and so can be utilised to support the growth process of the plant.

Leguminous plants and species of Rhizobium The group of flowering plants belonging to the Leguminosae family are found in both temperate and tropical climates and include peas, beans, clover, lupin, soybean and peanut. The majority of the members of this family of plants are capable of fixing symbiotically with bacteria of the genus *Rhizobium*, and this symbiotic association has proved extremely useful in agriculture. A certain degree of specificity is exhibited between the leguminous plant and the species of *Rhizobium* which will colonise it (Table 1.2). Rhizobia are to be found in nearly all soils but do not normally fix nitrogen in the free-living state. The bacteria colonise the roots of the plant at about the stage in the plant's growth when the first leaves begin to appear, although this colonisation process can differ in both nature and time of onset in different symbiotic associations.

The first step in the colonisation of the plant by the rhizobial bacteria involves the recognition of the host plant by the bacteria. Lectins are plant proteins which have been implicated as components of cell recognition systems and which appear to have binding sites with a specificity for particular sugar residues, thus being able to impart cohesiveness to cell surfaces by linking polysaccharide groups from two adjacent cells. These plant lectins may interact with specific rhizobia enabling the plant to recognise and admit the correct type of *Rhizobium*. The rhizobial bacteria enter the host through the root hair which becomes deformed due to a hormone-like substance produced by the bacteria. After entry into the root hair the bacteria migrate in a thread-like fashion, multiplying as they spread along the hair into the root tissue. Since different strains of *Rhizobium* differ in their ability to fix nitrogen depending upon the host plant which they infect, then it can be seen that it

Table 1.2   Host preference of *Rhizobium* species

Species	Preferred host genus
Rhizobium leguminosarum	*Pisum, Vicia, Lathyrus, Lens*
R. trifolii	*Trifolium*
R. phaseoli	*Phaseolus*
R. meliloti	*Medicago, Melilotus, Trigonella*
R. lupini	*Lupinus, Ornithopus*
R. japonicum	*Glycine max*
'Cowpea type'	*Vigna, Macroptillium* and others

Note: species *R. leguminosarum, trifolii, phaseoli* and *meliloti* are fast growers while *R. lupini* and *japonicum* are generally slow growers. The 'cowpea' rhizobia group contains a diverse range of *Rhizobium* which cannot be accommodated in the other groups and includes both slow- and fast-growing strains, some of which may even infect non-leguminous angiosperms.

is of significant agricultural importance that the rhizobia which infect a crop are effective in nitrogen fixation, especially since once the plant is colonised by a certain strain of *Rhizobium*, other strains are excluded.

Colonisation of root cells in the host plant tissue occurs when bacteria are released from the infection thread and involves the action of two cell wall hydrolysing enzymes, a pectinase from *Rhizobium* and a cellulase from the plant cell to facilitate entry of the bacteria into the plant cell where the bacteria continue to divide. As a consequence of this cell division, the plant cell becomes engorged with bacteria which then cease growing, enlarge up to forty-fold, and simultaneously undergo modification in both structure and function to become bacteroids which are rich in nitrogenase, the enzyme responsible for nitrogen fixation. Concurrent with the invasion of the host plant cell by rhizobia, there occurs a rapid burst of cell division involving both the invaded host cell and several layers of non-invaded neighbouring cells, which assists in the dissemination of the rhizobia and also produces the characteristic root nodules found in this symbiotic association.

Within this nodule an inner region of invaded and non-invaded cells is surrounded by an outer cortex of non-invaded host cells. In the invaded cells of this inner region of the mature root nodule, depending on the host-infecting rhizobial strain combination, one or more bacteroids is enclosed within a membranous envelope which is the probable site of occurrence of a red pigmented protein, leghaemoglobin, which is responsible for the characteristic colouration of nodules actively fixing nitrogen. Leghaemoglobin is a protein which is synthesised exclusively in the nitrogen-fixing root nodules, and is restricted to the infected cells within these nodules where it constitutes 25–30 per cent of the total soluble protein of the cell. Leghaemoglobin has myoglobin-like properties (myoglobin is a muscle protein whose function is thought to be that of assisting diffusion of oxygen into tissues and possibly providing a 'store' of oxygen), and its synthesis is coded for by the DNA of the host plant cell genome and not by the DNA of the infecting *Rhizobium*. The function of leghaemoglobin in the process of nitrogen fixation will be discussed later.

Before this symbiotic system is capable of fixing nitrogen three essential requirements must be met:

1 establishment of a root nodule;
2 differentiation of bacteria into bacteroids;
3 production of leghaemoglobin.

Providing that these conditions have been fulfilled then nitrogen fixation will continue throughout the life cycle of the plant, usually until seed formation when the nodules become senescent and nitrogen fixation ceases. Some of the bacteroids will be incapable of further multiplication but some will remain viable and, after ageing or death of the plant, will survive in the soil for the next season's growth.

Nodulated non-legumes   A number of non-leguminous plants, principally trees and shrubs, including those of the genera *Alnus* (the most intensively studied), *Ceanothus* and *Myrica* have root nodules in which the invading bacteria are not rhizobia. These plants have a wide geographical distribution and although much less detail is known about this nitrogen-fixing system, the input of nitrogen on a global scale must be enormous. Some non-leguminous angiosperms which form root nodules can fix atmospheric nitrogen at rates comparable to legumes and they contribute major amounts of fixed nitrogen to forests, wetlands, fields and other natural environments in which they grow. Until recently the symbiotic soil organism which infected the root cells to produce nodules had proved difficult to isolate but recently a soil actinomycete has been isolated which is capable of producing nitrogen-fixing symbiotic nodules on the roots of the woody angiosperm *Comptonia peregrina* (sweet fern). This class of bacteria, the Actinomycetes, had for several years been suspected to be the symbiotic nitrogen-fixing microbe. There is no evidence for the presence of leghaemoglobin in the root nodules of this symbiotic system although the nodules are pigmented, but it is likely that a protein with similar properties must be present since the nitrogenase from this system shows the characteristic oxygen sensitivity of that isolated from other well-studied nitrogen-fixing organisms. One exception to the general observation that *Rhizobium* only colonises the roots of leguminous plants is the discovery that a cowpea *Rhizobium* is capable of infecting woody non-leguminous plants of the genus *Trema*.

Algal symbioses   These symbioses result from the association of a blue-green alga with a plant. However, in contrast to Rhizobia and Actinomycetes where the associations are formed only with the most advanced plants, the dicotyledonous Angiosperms, the blue-green algae tend to favour the more primitive members of the plant kingdom, e.g. lichens, liverworts and pteridophytes. The only exception to this is the association between *Nostoc* an algal species which fixes nitrogen in heterocysts in this symbiotic association and *Gennera* a subtropical plant where nodules are formed at the base of the leaves. The nitrogen fixed within these nodules can then be utilised by the plant for metabolism and growth. Little is known about the ecological importance and biochemistry of these symbiotic systems.

Associative symbioses This term describes a vague grouping of symbiotic systems in which there is some interdependence between the partners although both may grow satisfactorily apart. Nitrogen-fixing prokaryotes can occur in association with the leaves and roots of higher plants, particularly in the tropics and although the original figures for the amount of nitrogen fixed by the associations between bacteria (species of *Spirillum*) and tropical grasses may be somewhat overestimated this area of research into nitrogen fixation may yet produce some useful efficient nitrogen-fixing associative symbioses between bacteria and higher plants.

The biochemistry of nitrogen fixation

In the industrial production of nitrogenous fertiliser, atmospheric dinitrogen is catalytically reduced to ammonia by reaction with hydrogen (produced from natural gas) in the Haber–Bosch process:

$$N_2 + 3H_2 \rightarrow 2NH_3$$

High pressures and moderately high temperatures are required for the reaction to be an efficient means of producing ammonia. How then can this reaction which requires quite extreme conditions when performed chemically be undertaken by living cells growing at low temperatures and atmospheric pressure?

Biological nitrogen fixation is approximately twice as efficient as the Haber–Bosch process but still has an energy requirement of 355 kJ mol^{-1} NH$_4{}^+$, this input of energy being required to overcome the activation energy for the reduction of dinitrogen. The nitrogenase enzyme system which performs this reduction process is unique to nitrogen-fixing organisms and the reaction catalysed can be represented as the six-electron reduction of dinitrogen to ammonia:

$$N_2 + 6H^+ + 6e^- \rightarrow 2NH_3$$

Nitrogenase is also able to reduce many other molecules which resemble dinitrogen in that they contain a triple bond and the requirements for the reduction of these compounds are identical to those for nitrogen reduction. Some examples are listed below:

1 Two-electron reduction processes:
   (a) acetylene → ethylene: $C_2H_2 + 2H^+ + 2e^- \rightarrow C_2H_4$
   (b) protons → hydrogen: $2H^+ + 2e^- \rightarrow H_2$
   (c) azide → nitrogen and ammonia: $N_3^- + 4H^+ + 2e^- \rightarrow N_2 + NH_4^+$

2 Six-electron reduction processes:
   cyanide → methane and ammonia: $CN^- + 8H^+ + 6e^- \rightarrow CH_4 + NH_4^+$

The ability of nitrogenase to reduce acetylene has proved to be of great practical significance in nitrogen fixation studies since the product, ethylene, can be detected rapidly and with great sensitivity by gas chromatography and is now the basis of the acetylene test for nitrogen fixation. This method gives results which are quantitative since the levels of acetylene reduction and nitrogen fixation are directly related and the development of this test has permitted the easy and reliable measurement of the nitrogen-fixing ability of cells. The basic cellular requirements for the biological reduction of nitrogen are listed below:

1 an active nitrogenase enzyme;
2 Mg^{2+} and a continuous supply of ATP;
3 a strong reducing agent, i.e. a suitable electron donor;
4 low oxygen tension.

Nitrogenase   The enzyme isolated from several different sources has proved to have extremely similar properties and always consists of two oxygen-sensitive non-haem iron proteins. During purification oxygen must be excluded or the enzyme will be inactivated and this oxygen sensitivity has necessitated the use of a technique known as anaerobic chromatography in the purification of nitrogenase. The enzyme can be separated into two distinct proteins both of which are brown in colour due to the presence of iron and neither of which is active in nitrogen fixation unless the other is present. The larger protein, the (Mo–Fe)-protein, consists of four subunits of two different types and contains molybdenum in addition to iron, while the smaller Fe-containing protein consists of two identical subunits and contains no molybdenum.

Table 1.3 Some properties of nitrogenase from *Klebsiella pneumoniae*

	Mo – Fe protein	Fe – protein
Molecular weight	218,000	66,800
Number of subunits	Two types	Two identical subunits
	2 × 59,000	34,000
	2 × 51,000	
Mo content, g-atom/mol	2	none
Fe content, g-atom/mol	24–36	4
S^{2-} content, g-atom/mol	Not determined	4

Some properties of the enzyme isolated from *Klebsiella pneumoniae* are listed in Table 1.3.

The iron atoms are usually accompanied by essentially the same number of sulphur atoms which are attached to the iron atoms, and the smaller protein is especially susceptible to oxygen inactivation. The first stable product of nitrogen fixation is ammonia and, while it has not been possible to detect any other free intermediates of the reductive process, e.g. diimide (N_2H_2) or hydrazine (N_2H_4), it has been possible to demonstrate the presence of an enzyme-bound dinitrogen hydride intermediate which is only present when the enzyme is reducing nitrogen.

ATP requirements During nitrogen fixation ATP, or to be more precise, the mono-magnesium salt of ATP, is hydrolysed to the mono-magnesium salt of ADP and inorganic phosphate. Similarly the reduction of all other substrates by nitrogenase is coupled to the hydrolysis of ATP. Calculations of the amount of ATP utilised by the cell in the biological reduction of nitrogen are complicated by the fact that ATP is utilised and hydrolysed in many other cellular processes in addition to nitrogen fixation. However, *in vitro* studies give estimates of 12–15 moles of ATP consumed per mole of nitrogen reduced to ammonia, i.e. 4–5 moles of ATP are required per electron pair transferred from reductant to nitrogen in the reduction process. The ATP required for nitrogen fixation, which in physiological terms is a bioenergetically expensive process, can be supplied by various means, e.g.

1 Oxidation of respiratory substrates (in aerobic organisms)
2 In the anaerobic *Clostridium* via the 'phosphoroclastic' cleavage of pyruvate to acetate, since no external oxidant

is needed in this reaction:

$$\text{Pyruvate} \xrightarrow[\substack{\text{Thiamine pyrophosphate} \\ \text{Ferredoxin}}]{\text{Coenzyme A}} \text{Acetyl CoA} + \text{Reduced ferredoxin} + CO_2$$

$$\left.\begin{array}{l} \text{Acetyl CoA} + P_i \rightleftharpoons \text{Acetyl phosphate} + \text{CoASH} \\ \text{Acetyl phosphate} + \text{ADP} \rightleftharpoons \text{Acetate} + \text{ATP} \end{array}\right\} \text{ATP-generating system}$$

3 Photosynthetic systems – via photophosphorylation.

Source of reductant Ferredoxins and flavodoxins are small specialised proteins of molecular weights in the range of 6,000–24,000, which can exist in oxidised and reduced forms that are readily interconvertible. The reduced forms of these proteins are strong reducing agents which are capable of reducing many other biological molecules provided that the appropriate enzyme to catalyse the reaction is present.

Ferredoxins are involved in a variety of biological processes including photosynthesis (plants) and pyruvate metabolism (anaerobic bacteria). While the 'phosphoroclastic' cleavage of pyruvate may provide a general mechanism for the generation of reduced ferredoxin in *Clostridium*, where ferredoxin appears to be the actual protein which reacts with the nitrogenase enzyme and provides reducing power for the reduction of nitrogen, this 'phosphoroclastic' cleavage does not occur universally and other mechanisms of reductant generation need to be considered. For the aerobe *Azotobacter chroococcum* ferredoxin is present but a flavodoxin (azotoflavin) is the primary reductant for nitrogenase.

Flavodoxins, which are proteins containing a flavin radical, are yellow proteins when fully oxidised, colourless when fully reduced and blue when half reduced in the semiquinone state. Flavodoxins do not contain iron atoms and their oxido-reducible centre is the fluorescent flavin molecule (Fig. 1.2). The semiquinone form is unusually stable to oxidation by air and for this reason would be preferable to ferredoxin and more suited to the aerobic way of life of organisms such as *Azotobacter*.

There is still much uncertainty regarding the *in vivo* generation of electrons for the reduction process. Ferredoxin can be reduced by NADPH through the reversal of the reaction catalysed by NADP-ferredoxin reductase, thus if NADPH could be produced in the cell it could serve indirectly as an electron donor for nitrogen fixation. In *Azotobacter vinelandii*, a non-photosynthetic

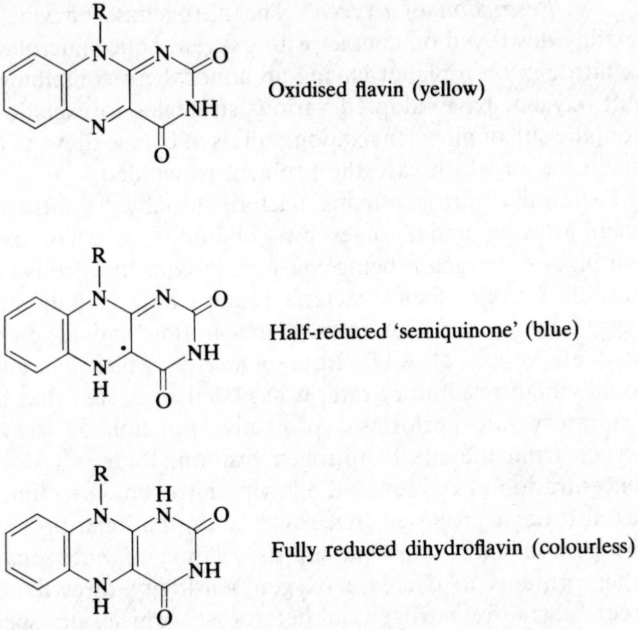

Oxidised flavin (yellow)

Half-reduced 'semiquinone' (blue)

Fully reduced dihydroflavin (colourless)

Fig. 1.2 Oxidised, half-reduced and fully reduced flavins

aerobic bacterium, NADPH appears to be generated by isocitrate dehydrogenase activity:

$$\text{Isocitrate} + \text{NADP}^+ \underset{\text{dehydrogenase}}{\overset{\text{Isocitrate}}{\rightleftharpoons}} \text{2-Oxoglutarate} + CO_2 + \text{NADPH} + H^+$$

Electrons for the reduction process can then be transferred through a specific azotobacter ferredoxin to azotoflavodoxin which is the immediate electron donor for nitrogenase in this bacterium:

$$\text{NADPH} \rightarrow \text{Ferredoxin} \rightarrow \text{Azotoflavin} \rightarrow \underset{\text{(Nitrogenase)}}{\text{Fe-protein}} \rightarrow \underset{\text{(Nitrogenase)}}{\text{(Mo–Fe)-protein}} \rightarrow N_2$$

There have been reports of an NADP^+-specific isocitrate dehydrogenase in the bacteroids of root nodules in legumes. This may be of particular significance if NADPH is the electron donor for nitrogenase. A coincidence of rapid isocitrate dehydrogenase activity in pea nodule bacteroids with a peak of nitrogen fixation in the entire plant provides additional presumptive evidence that isocitrate dehydrogenase may provide NADPH for the fixation process.

Restriction of oxygen The nitrogenase enzyme is irreversibly destroyed on contact with oxygen, hence microbes which fix nitrogen on a planet having an atmosphere containing 20 per cent oxygen, have adopted various strategies to exclude oxygen from the site of nitrogen fixation, unless of course they are obligate anaerobes in which case the problem is avoided.

Facultative nitrogen-fixing bacteria usually fix nitrogen only when growing under anaerobic conditions or when oxygen is limiting, i.e. oxygen is being consumed as quickly as it is supplied. Aerobic nitrogen-fixing bacteria require an adequate supply of oxygen to sustain high rates of respiration and to provide an adequate supply of ATP. In *Azotobacter*, which has an exceptionally high respiratory rate, it has been suggested that the high respiratory rate performs a protective function by 'scavenging' oxygen from the site of nitrogen fixation, thereby reducing the concentration of oxygen and allowing nitrogenase to function. It has also been proposed that some conformational protection of nitrogenase may occur via the association of nitrogenase with other proteins to decrease oxygen sensitivity. Free-living blue-green algae fix nitrogen in heterocysts which are specialised cells occurring at regular intervals along the filaments. The heterocyst wall prevents oxygen entry into the cell and nitrogen fixation is usually restricted to the heterocysts while being absent from the vegetative filament.

The symbiotic nitrogen-fixing system found in root nodules has evolved a particularly elaborate system which utilises a special form of haemoglobin – leghaemoglobin – synthesised by the host plant to regulate oxygen concentration at the nitrogen-fixing site (see p. 7). Leghaemoglobin has such a high affinity for oxygen that it delivers oxygen to the rhizobia at a concentration which is harmless to their nitrogenase, so preventing the accumulation of high concentrations of free oxygen but simultaneously providing oxygen for rhizobial respiratory metabolism. Thus the root nodule may be considered as a highly specialised compartment providing an environment in which the fixation of nitrogen and oxidative metabolism are physiologically compatible.

ATP-dependent hydrogen evolution Under optimum conditions for nitrogen fixation *in vitro* only 75 per cent of the available electrons for nitrogen reduction are actually used for this purpose. The remaining 25 per cent are used to reduce protons to hydrogen in an ATP-dependent process catalysed by nitrogenase and requiring Mg^{2+} and reductant just as in the nitrogen reduction process.

Many nitrogen-fixing bacteria seem unable to prevent this hydrogen-yielding side reaction and this may account for the low 'efficiency' of nitrogen fixation in these cells, i.e. the ATP used in proton reduction, which can be almost one-third of the energy flow through nitrogenase, is effectively 'wasted'. *Azotobacter* evolves no hydrogen under normal circumstances since the hydrogenase of the cell (the enzyme normally catalysing both hydrogen evolution and hydrogen uptake in the cell, i.e. not that via nitrogenase) is not reversible and catalyses only hydrogen uptake, thus the hydrogen is effectively trapped by the cell and may be recycled.

Symbiotic nodule bacteria also evolve hydrogen to varying extents and in some plant–bacteria associations as much as 50 per cent of the available ATP and reducing power can be lost in hydrogen evolution. Indeed the loss of energy as ATP and reducing power in soybean and other legume root nodules which can be attributed to nitrogenase-catalysed hydrogen evolution may be as high as 40–60 per cent of the total energy flow through nitrogenase. The importance of this factor from the agricultural viewpoint is that, in terms of crop productivity, the 'tighter' (non-hydrogen-evolving) symbiotic systems are more efficient in that they are able to fix more nitrogen per unit of solar energy than the 'loose' hydrogen-evolving systems. In general, it appears that non-leguminous associations are 'tighter' than most leguminous associations.

Functions of the two protein components of nitrogenase
For nitrogenase to function, a complex between the Fe-protein and the (Mo–Fe)-protein must be formed with the two components present in a 1:1 molar ratio. The (Mo–Fe)-protein can exist in three different states, each state being characterised by the proportion of iron atoms in the reduced (ferrous) state in the molecule. The most oxidised form appears to have no physiological importance and it is in the intermediate form that the protein is usually isolated. The most reduced form only appears when all the conditions for nitrogen fixation are satisfied, i.e. when ATP, Mg^{2+} and reductant are present, and this is the predominant form when the enzyme is functioning in nitrogen fixation.

Although the precise mechanism by which nitrogenase fixes nitrogen has yet to be elucidated it is thought that the

(Mo–Fe)-protein reacts with the potential substrates for nitrogenase, e.g. N_2, C_2H_2, probably via interaction of the reducible substrate with the molybdenum atoms of the protein. The smaller Fe-protein binds ATP as the mono-magnesium salt and becomes a more powerful reducing agent by receiving electrons from the physiological electron donor system and in this reduced state is now capable of generating the most reduced form of the (Mo–Fe)-protein. In outline, the fixation process involves the reduction of Fe atoms in the smaller Fe-protein and as a consequence ATP is hydrolysed to ADP and P_i. An electron from the Fe atoms in the (Mo–Fe)-protein is now utilised to reduce the substrate bound to the (Mo–Fe)-protein. Several of these electron transfer steps will be required before the final reduced product (NH_3 in the case of N_2 substrate) is released from the enzyme. Each time an electron transfer between the Fe-protein and the (Mo–Fe)-protein takes place, ATP is hydrolysed to ADP. The overall reduction process is illustrated schematically in Fig. 1.3.

Control of nitrogenase   Nitrogenase can be considered an expensive enzyme in terms of biological energy consumption since it requires a constant and plentiful supply of ATP for continued activity, consequently nitrogen-fixing organisms have developed a means of regulating both the activity and the synthesis of nitrogenase. The capacity to fix nitrogen is governed by a set of genes

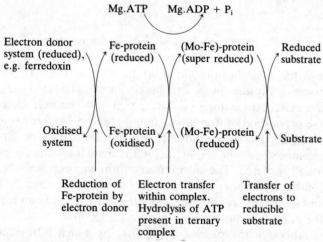

Fig. 1.3 Reduction of substrates by nitrogenase

called the nitrogen fixation (*nif*) genes. Free-living bacteria couple NH_4^+ production by nitrogenase to cellular biosynthesis and do not export any detectable NH_4^+ into the external environment. When combined nitrogen, e.g. NH_4^+ or NO_3^- is present in the environment as a nitrogen source for growth then the *nif* genes of these organisms are repressed and no nitrogenase is produced.

Conversely, the *nif* genes of symbiotic bacteria such as rhizobia are normally 'derepressed' leading to the synthesis and export of fixed nitrogen to the external environment (the cytoplasm of the host cell in the case of *Rhizobium*), as nitrogenase is produced and is active even in the presence of ammonia. The derepression of *nif* genes in this symbiotic association results in the production of large quantities of fixed nitrogen from atmospheric nitrogen and has a crucial role to play in man's food chain on earth. Nitrogen fixation in this symbiotic system may be subject to control via the supply of energy (ATP) which could become a rate-limiting step.

Relatively few details concerning regulation of the activity of nitrogenase are known. It has been shown from *in vitro* studies that ATP is essential for enzyme activity and that ADP, the product of ATP utilisation by the enzymes, inhibits nitrogenase activity. Direct control of nitrogenase activity may be mediated through the ATP:ADP ratio in the vicinity of the enzyme thereby regulating the rate at which the enzyme reduces substrate. Indirect control of enzyme activity may be exerted via the partitioning of electrons between the linked processes of nitrogen reduction and ATP-dependent hydrogen evolution, i.e. the rate of hydrogen evolution may increase relative to the rate of nitrogen fixation, thereby removing 'excess' reducing power particularly if hydrogenase activity, concerned with the 'recapturing' of electrons by the cell, decreases.

Ammonia, the primary product of nitrogenase activity, suppresses the biosynthesis of nitrogenase in free-living bacteria and in *Azotobacter* and *Klebsiella* it has been shown that the synthesis of both nitrogenase protein components is regulated in a coordinated fashion. There is strong evidence for the regulation of transcription of nitrogenase genes via glutamine synthetase, the enzyme responsible for the assimilation of NH_4^+ into glutamine in the cell:

$$\underset{\text{Glutamic acid}}{\overset{NH_4^+}{\Big\downarrow}} \xrightarrow[\underset{\text{ADP} + P_i}{\text{ATP}}]{\text{Glutamine synthetase}} \text{Glutamine}$$

Glutamine synthetase exists in these organisms in an active (deadenylylated) or an inactive (adenylylated) form. When ammonia levels are high, glutamine synthetase is adenylylated by the cell and thus inactivated with a consequential switching off of the transcription of the *nif* genes. When ammonia levels are low glutamine synthetase becomes deadenylylated and thus activated resulting in the promotion of transcription of the nitrogenase genes and, as a consequence, nitrogenase biosynthesis and enhanced nitrogen fixation. In both *Klebsiella* and *Rhizobium* the level of adenylylated (inactive) glutamine synthetase is inversely related to the level of nitrogenase enzyme in the cell.

The control of glutamine synthetase activity via adenylylation of the enzyme (covalent attachment of 5′-AMP residues to tyrosyl hydroxyls on the glutamine synthetase subunits) is mediated through the activity of adenylyltransferase (ATase) which catalyses both the adenylylation and deadenylylation reactions (Fig. 1.4).

Unless the adenylylation and deadenylylation reactions are appropriately controlled a futile cycle will occur resulting in the fluctuation of glutamine synthetases between the active and inactive forms, while ATP is hydrolysed to ADP and P_i. This coupling is prevented by the action of a regulatory protein, P_{II}, which is present as a complex with ATase and can itself exist in two interconvertible forms. As seen in Fig. 1.4, P_{II} stimulates the adenylylation reaction, but P_{II} can be uridylylated (covalent attachment of UMP moiety to the protein) by the enzyme uridylyl transferase (UTase) and this uridylylated form stimulates the deadenylylation reaction. Uridylylation of P_{II} and hence the

(i) $GS + ATP \xrightarrow[P_{II}]{ATase} GS - AMP + PP_i$ adenylylation, GS inactive

(ii) $GS - AMP + P_i \xrightarrow[P_{II}-UMP]{ATase} GS + ADP$ deadenylylation, GS active

(iii) $P_{II} + UTP \xrightarrow{UTase} P_{II} - UMP + PP_i$ uridylylation

(iv) $P_{II} - UMP + H_2O \xrightarrow{UR\ enzyme} P_{II} + UMP$ deuridylylation

GS   = glutamine synthetase
ATase   = adenylyl transferase
P_{II} = regulatory protein
$P_{II} - UMP$ = uridylylated form of regulatory protein
UTase   = uridylyl transferase
UR enzyme = enzyme responsible for removal of uridylyl group from $P_{II} - UMP$

Fig. 1.4 Interconversions of active and inactive forms of
glutamine synthetase

subsequent production of the activated form of glutamine synthetase is stimulated by a decrease in the cellular level of NH_4^+ and an increase in the oxoglutarate:glutamine ratio in the cell. The consequence of this stimulation of the deadenylylation reaction in conditions where the cell requires an increased fixed-nitrogen supply is the derepression of the *nif* genes and subsequent nitrogenase production in the cell, thus allowing the cell to increase the rate of nitrogen fixation. Conversely, the uridylylation of P_{II} to give P_{II}-UMP is inhibited by high levels of NH_4^+ in the cell and by low oxoglutarate:glutamine ratios, and under these conditions there is an increase in the proportion of inactivated glutamine synthetase in the cell, a consequential 'switching off' of the *nif* genes and ultimately a reduction in the level of nitrogenase activity and nitrogen fixation in the cell.

The control of nitrogenase biosynthesis by glutamine synthetase is summarised in Fig. 1.5. These factors are not the only cellular controls regulating the expression of the *nif* genes in free-living nitrogen-fixers, such as *Klebsiella*. Oxygen is also known to repress nitrogenase biosynthesis, but by a mechanism which is not mediated via glutamine synthetase.

Nitrification

Ammonia which has been released into the soil as a result of the processes of nitrogen fixation or ammonification can be taken up directly by plants or oxidised to nitrite and nitrate. This oxidation process, known as nitrification, is performed by two highly specialised groups of soil bacteria known collectively as nitrifying bacteria. These bacteria cannot use external sources of organic matter for their growth (i.e. they are autotrophs), but they use the 'energy' derived from the oxidation of ammonia to 'fix' carbon dioxide into organic linkage.

Nitrosomonas is typical of several genera which possess the ability to oxidise ammonia to nitrate:

$$NH_4^+ + 1.5O_2 \rightarrow NO_2^- + 2H^+ + H_2O$$

This oxidation appears to be a complex process which occurs in three stages via the intermediates hydroxylamine and nitroxyl:

$$NH_4^+ \xrightarrow[\substack{3H^+ \\ 2e^-}]{H_2O} NH_2OH \xrightarrow[\substack{2H^+ \\ 2e^-}]{} NOH \xrightarrow[\substack{3H^+ \\ 2e^-}]{H_2O} NO_2^-$$

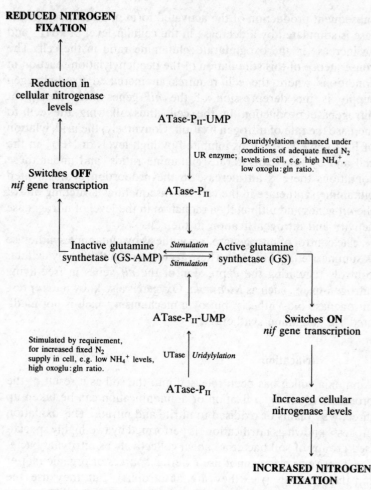

Fig. 1.5 Control of nitrogenase biosynthesis by glutamine synthetase

Although the details of the oxidation steps have yet to be elucidated, the oxidation of hydroxylamine to nitrite, at least, appears to involve components of the cytochrome electron transport chain.

Nitrobacter is a genus of soil bacteria which possesses the capability of oxidising nitrite to nitrate:

$$NO_2^- + \tfrac{1}{2}O_2 \rightarrow NO_3^-$$

The oxidation of nitrite to nitrate appears to be mediated via a cytochrome electron transport chain in the membranes of *Nitrobacter*. Nitrite entering the bacterial cell is oxidised on several double-layered membranes which completely envelop the interior of the cell and the inorganic ions are thus prevented from penetrating to the cell interior where they may have toxic effects. The true substrate in nitrite oxidation appears to be a hydrated form of nitrite and oxidation takes place utilising an oxygen atom derived not from atmospheric oxygen, but from water. Some of the ATP which is generated from the passage of electrons from nitrite to oxygen is utilised in the production of reduced pyridine nucleotides by reversing the flow of electrons along the electron transport chain. These reduced pyridine nucleotides may then be utilised in biosynthetic reactions in the cell.

Through the combined actions of the nitrifying bacteria, ammonia present in the soil can be oxidised to nitrate. Ammonia itself is an adequate plant nutrient but most plants prefer nitrate, a more effective form of nitrogen in plant nutrition. In this process of nitrification, the nitrifying bacteria have thus made the ammonia-N more readily available to the plant. One disadvantage of the nitrification process is that whereas ammonia is well retained by soil, nitrate is readily washed away, particularly in the surface layers of the soil where the roots of plants are to be found. The process of nitrification can thus become of economic significance when artificial fertilisers are used. Application of free ammonia or ammonia salt fertilisers can be rendered less efficient if the ammonia is converted to nitrate as this can lead to increased loss of nitrogen from the soil and contamination of the surrounding waters. This may be prevented to some extent by the use of agricultural chemicals which restrict 'run-off' of artificial fertiliser by inhibiting the growth of nitrifying bacteria.

Denitrification

Nitrate in the soil may be taken up by higher plants and further metabolised or used as an alternative oxidant to oxygen by bacteria and thereby lost from the soil to the atmosphere as nitrous oxide or dinitrogen in a process known as denitrification. Denitrification occurs most readily in anaerobic conditions where the bacteria use NO_3^- or NO_2^- as terminal hydrogen acceptor in an energy-yielding reaction in place of unavailable O_2. Nitrate may initially be reduced to nitrite:

$$NO_3^- + 2H^+ + 2e^- \rightarrow NO_2^- + H_2O$$

This reduction step is catalysed by the nitrate reductase enzyme in the bacteria. The enzyme is tightly bound to the bacterial membrane and contains molybdenum atoms which are thought to interact directly with the nitrate.

There exist several bacteria which are capable of reducing nitrate or nitrite to dinitrogen (or in some cases N_2O), and these can be found commonly in sewage treatment plants and compost heaps. Examples of such denitrifying bacteria are to be found in the genera *Pseudomonas*, *Micrococcus* and *Thiobacillus*. The gaseous product of this process thus becomes unavailable for use by plants and represents a loss of biological fixed nitrogen from the soil. The enzymology of this denitrification process is not clear, but nitrate reduction may take place in a stepwise manner.

Nitrate assimilation

The assimilation of nitrate in higher plants and algae occurs in three stages:

1 nitrate entry into plant cells;
2 the reduction of nitrate to nitrite;
3 the reduction of nitrite to ammonia.

The plasma membrane of the cell forms a barrier to the uptake of nitrate and the initial uptake of nitrate is an inducible process involving the appearance of a nitrate-specific permease in the membrane. Once nitrate has entered the cell it is reduced to ammonia by two different enzymes, nitrate reductase which reduces nitrate to nitrite, a stable free intermediate, and nitrite reductase which reduces the nitrite further to the stable final product, ammonia. The reduction steps require the donation of eight electrons per molecule of NH_4^+ formed and utilise NADH and reduced ferredoxin as electron donors in the process which can occur in both plant roots and leaves.

Nitrate reductase

Nitrate reductase is the enzyme which is responsible for the reduction of nitrate to nitrite in the plant cell, and is a complex enzyme having a high molecular weight (up to 500,000 depending on the source). It also contains several prosthetic groups including FAD and molybdenum, although the precise role of these components in the process of nitrate reduction remains unclear. The enzyme appears to be localised mainly in the soluble portion of the cell and in the majority of cases the enzyme from higher plants demonstrates a specific requirement for NADH as electron donor for the reduction process. During the enzymic conversion of nitrate to nitrite by nitrate reductase, the substrate, i.e. nitrate, is thought to become directly attached to a molybdenum centre of the enzyme prior to being reduced. In the overall reduction process nitrate is reduced via nitrite to ammonia:

$$NO_3^- \xrightarrow{\text{Reduction}} NO_2^- \xrightarrow{\text{Reduction}} NH_3$$

It has been suggested that the role of the molybdenum centre in the nitrate reductase enzyme is to allow the removal of only one oxygen atom from the nitrate group and so avoid the production of NO, N_2O, N_2, i.e. denitrification. When the oxygen atom has been removed from nitrate, the nitrite formed can be bound to nitrite reductase via its nitrogen atom and the reduction process then completed without loss of any nitrogen.

Nitrate reductase is a key enzyme in the regulation of the assimilatory reduction of nitrate in higher plants. Nitrate reductase activity can be induced in higher plant cells by the presence of the substrate, nitrate, and the enzyme can exist in two stable forms one of which is active in nitrate reduction, while the other is inactive. The two forms may be interconverted by an oxidation – reduction mechanism and under reducing conditions the active form is converted into an inactive form. Indeed, nitrate reductase activity may be controlled to some extent by the internal redox state of the cell, an increase in the level of reduced pyridine nucleotides, i.e. 'reducing power', may inactivate the enzyme. In green algae it has been suggested that the nitrate reductase activity may be regulated by the level of ammonia which, when added to growth medium which also contains nitrate, can promote the inactivation of nitrate reductase in the cell and eventually repress the synthesis of nitrate reductase. The action of ammonia may be mediated via alteration of the redox state of the cell, since the presence of ammonia can promote an increase in the level of reduced pyridine nucleotides in growing algae.

There also appears to be a close relationship between photosynthesis and nitrate metabolism, and at low light intensities nitrate has been shown to accumulate in the leaves of higher plants. The reason for this seems to be a reduction in the level of reduced pyridine nucleotides in the cytoplasm of leaf cells arising from a reduction in the supply of glycolytic intermediates released from the chloroplast into the cytoplasm as a result of reduced photosynthetic fixation of CO_2 in the chloroplast. The oxidation of these glycolytic intermediates (e.g. glyceraldehyde-3-phosphate) would normally produce NADH in the cytoplasm which could then be utilised in cytoplasmic reduction processes including nitrate reduction.

Nitrite reductase

The reduction of nitrite to ammonia is catalysed by one enzyme, nitrite reductase, which in green plants has a relatively low molecular weight (about 62,000) and possesses an unusual haem compound – sirohaem – which is a tetrahydroporphyrin having adjacent pyrrole rings reduced and possessing eight carboxylate side chains (Fig. 1.6).

The reduction reaction catalysed by nitrite reductase involves the transfer of six electrons to the substrate nitrite:

$$NO_2^- + 6e^- + 8H^+ \rightarrow NH_4^+ + 2H_2O$$

The substrate nitrite, produced by the action of nitrate reductase on nitrate, rarely accumulates under normal conditions since nitrite reductase is invariably present at much higher levels of activity than nitrate reductase. There appears to be no free intermediate released during the reduction process, which may proceed via the formation of an NO haem intermediate, representing a one-electron reduction step which would then require another five electrons to be added to complete the reduction. However, the precise mechanism by which these reduction steps

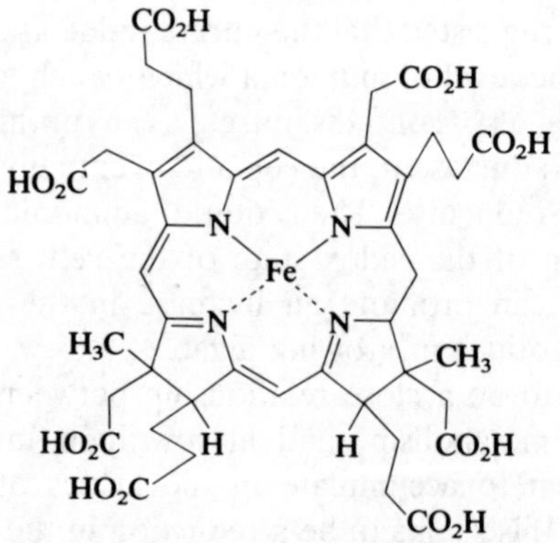

Fig. 1.6 Sirohaem

occur has not been elucidated. In plants, nitrite reductase receives electrons from a low potential electron donor, most probably ferredoxin, these electrons being ultimately accepted by the sirohaem molecule which is also the site of attachment of the substrate, nitrite, and most probably of the other intermediates in the reduction steps leading to ammonium production.

In leaves, nitrite reductase is found in chloroplasts where the electron donor, reduced ferredoxin, can be generated via coupling to cyclic photosynthetic electron transport. Nitrite is also readily metabolised by roots where nitrite reductase is associated with proplastids but, in the absence of ferredoxin from root tissue, an electron donor capable of coupling the transfer of electrons from NADPH (a possible electron donor produced by the enzymes of the pentose phosphate pathway in the proplastids) to the nitrite reductase has not yet been isolated.

Ammonia dissimilation

The plant or microbe growing in the soil assimilates inorganic nitrogen mainly in the form of nitrate, which is then reduced to ammonia intracellularly prior to incorporation into organic linkage primarily via the biosynthesis of amino acids. The three enzymes which have been considered to be of importance in the assimilation of ammonia and the reactions which they catalyse are listed below:

1 Glutamate dehydrogenase
2-oxoglutarate $+ NH_3 + NAD(P)H + H^+ \rightleftharpoons$
L-glutamate $+ NAD(P)^+ + H_2O$

2 Glutamine synthetase

$$\text{L-glutamate} + NH_3 + ATP \xrightleftharpoons{\text{Mg}^{2+}\text{ or Mn}^{2+}} \text{L-glutamine} + ADP + P_i$$

3 Glutamate synthase (L-glutamate:$NADP^+$ oxidoreductase (transaminating); trivial name GOGAT)
   (a) 2-oxoglutarate $+$ L-glutamine $+$ reduced ferredoxin $\rightleftharpoons$
   2 L-glutamate $+$ oxidised ferredoxin
   (b) 2-oxoglutarate $+$ L-glutamine $+ NAD(P)H + H^+ \rightleftharpoons$
   2 L-glutamate $+ NAD(P)^+$

Glutamate dehydrogenase

Glutamate dehydrogenase acting in an aminating role could catalyse the assimilation of ammonia into organic linkage through the formation of the amino acid glutamic acid from 2-oxoglutarate.

The enzyme from pea roots has a molecular weight of 208,000 and requires a divalent metal ion for activity in the aminating direction. The enzyme is localised principally in mitochondria in both the roots and leaves of higher plants, although low levels of the enzyme may also be found in chloroplasts. Although it is theoretically possible for glutamate dehydrogenase to play a crucial role in ammonia assimilation, there are two important factors opposing this. Firstly, the mitochondrial location of the enzyme would suggest that glutamate dehydrogenase was functioning in a degradative fashion since mitochondria normally oxidise glutamic acid under physiological conditions. Secondly, and perhaps more significantly, kinetic studies have demonstrated that glutamate dehydrogenase has a low affinity for ammonia (K_m value in the range 5–100 mM) which is inconsistent with an assimilatory role for this enzyme in the cell, where the physiological concentration of ammonia tends to be much lower than the K_m value. It is now thought that the assimilation of ammonia into amino acids, under normal physiological conditions, occurs via the action of the enzymes glutamine synthetase and glutamate synthase (GOGAT).

Glutamine synthetase

Glutamine synthetase can be found in both the chloroplasts and cytoplasm in plant cells and catalyses the formation of glutamine from glutamate with the simultaneous cleavage of ATP to ADP. The enzyme from plant tissue has a molecular weight in the range 330,000–376,000 and the enzyme from soybean nodules consists of eight monomers of molecular weight 47,300 arranged in two parallel sets of planar tetramers. The pH optimum of the enzyme is dependent upon the divalent metal ion employed in the reaction, being about pH 8.0 when Mg^{2+} is used and about pH 5.0 with Mn^{2+} as the metal ion. The physiological significance of this observation is difficult to relate since the relative concentrations of Mg^{2+} and Mn^{2+} at the site of action of glutamine synthetase in the cell are not known. The activity of glutamine synthetase is affected by changes in Mg^{2+}, pH and energy charge, all of which are known to change in favour of increased glutamine synthetase activity in chloroplasts upon illumination. The glutamine synthetase in the cytosol of soybean nodules exhibits feedback inhibition by glutamine and an apparent regulation by energy charge. Apart from these observations, little is known about the control of glutamine synthetase activity in higher plants but adenylylation of glutamine synthetase does not appear to occur in higher plants in contrast to the situation in bacteria.

Glutamate synthase

The enzyme glutamate synthase (GOGAT) catalyses the reductive transfer of an amide-amino group from glutamine to 2-oxoglutarate with the resultant production of two molecules of glutamate. GOGAT has been identified in many higher plants and algae and either reduced ferredoxin or NAD(P)H is required for activity depending upon the source from which the enzyme is isolated. The ferredoxin-dependent enzyme from *Vicia faba* has a molecular weight of about 150,000 and in leaf tissue the ferredoxin-dependent enzyme has been demonstrated to be present in chloroplasts. Product feedback inhibition of the NADH-dependent GOGAT from lupin nodules has been demonstrated, the enzyme being inhibited by glutamate competitively with respect to 2-oxoglutarate and by NAD^+ competitively with respect to NADH.

It was the discovery of GOGAT in higher plants that changed the emphasis from a central role for glutamate dehydrogenase in ammonia assimilation to one involving a central role for glutamine and the concerted action of glutamine synthetase and GOGAT. The assimilation route involving these enzymes is summarised in Fig. 1.7.

Ammonia is usually present at low concentrations in the soil as it is rapidly oxidised to nitrate by soil bacteria. When nitrate or ammonia is taken up by the plant roots, the ammonia present is rapidly assimilated and in the *in vivo* situation, it appears that there is usually a low NH_4^+ concentration in the cell. Consideration of the relative K_m values for glutamate dehydrogenase and glutamine synthetase shows that glutamine synthetase has a much higher affinity for NH_4^+, thus it is unlikely that under the *in vivo* situation of low NH_4^+ availability that glutamate dehydrogenase has any significant role to play in ammonia assimilation and that assimilation takes place via the glutamine synthetase/GOGAT pathway.

Now that the assimilated N has been converted into an organic form, it can be utilised by the cell for further biosynthetic purposes or transported and stored in various forms until required. The metabolism of these compounds containing organically bound nitrogen is discussed in the following chapters.

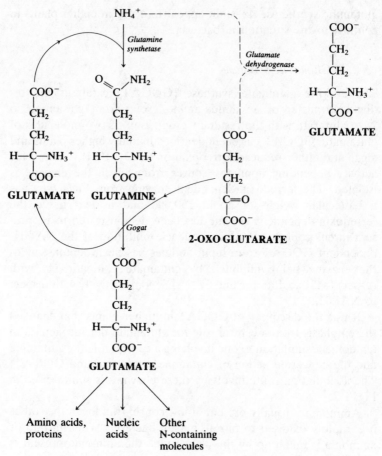

Fig. 1.7 Assimilation of ammonia by the actions of glutamine synthetase and GOGAT

Suggestions for further reading

Baulcombe, D. & Verma, D. P. S. (1978) Preparation of a complementary DNA for leghaemoglobin and direct demonstration that leghaemoglobin is encoded by the soybean genome, *Nucleic Acids Research*, **5**, 4141–53.

Callagham, D., Del Tredici, P. & Torrey, J. G. (1978) Isolation and cultivation *in vitro* of the Actinomycete causing root nodulation in *Comptonia*, *Science*, **199**, 899–902.

Dalton, M. & Mortensen, L. E. (1972) Dinitrogen (N_2) fixation (with a biochemical emphasis), *Bact. Rev.*, **36**, 231–60.

Postgate, J. (1978) *Nitrogen Fixation*, Studies in Biology no. 92. Edward Arnold: London.

Rawsthorne, S., Minchin, F. R., Summerfield, R. J., Cookson, C. & Coombs, J. (1980) Carbon and nitrogen metabolism in legume root nodules, *Phytochem.*, **19**, 341–55.

Segal, A., Brown, M. S. & Stadtman, E. R. (1974) Metabolite regulation of the state of adenylylation of glutamine synthetase, *Archives Biochem. Biophys.*, **161**, 319–27.

Shanmugan, K. T., O'Gara, F., Anderson, K. & Valentine, R. C. (1978) Biological nitrogen fixation, *Ann. Rev. Plant Physiol.*, **29**, 263–76.

Thorneley, R. N. F., Eady, R. R & Lowe, D. J. (1978) Biological nitrogen fixation by way of an enzyme-bound dinitrogen-hydride intermediate, *Nature (Lond.)*, **272**, 557–8.

2

Amino acid biosynthesis

The organically bound nitrogen of glutamic acid and glutamine can be utilised for the synthesis of other amino acids and subsequently be involved in the synthesis of proteins and other important nitrogen-containing molecules, e.g. purine and pyrimidine bases. Alternatively, the assimilated nitrogen may be transported to another part of the plant prior to subsequent metabolism, but whatever the fate of the assimilated nitrogen both glutamate and glutamine have been shown to play a pivotal role in these metabolic interconversions. A list of amino acids which are used by the cell for the biosynthesis of proteins is illustrated in Fig. 2.1

Although many of the details of the individual reactions have yet to be established, there is sound evidence that amino acid biosynthesis in plants follows the same reaction pattern found in micro-organisms. The carbon skeletons of the 20 amino acids utilised in protein synthesis are derived mainly from the intermediates of glycolysis, the tricarboxylic acid cycle and the pentose phosphate pathway. The six intermediates playing a major role in the supply of these carbon skeletons for amino acid biosynthesis are 2-oxoglutarate, oxaloacetate, pyruvate, 3-phosphoglycerate, phosphoenolpyruvate and erythrose-4-phosphate, (Fig. 2.2). Utilising these carbon skeletons as acceptor molecules and the compounds into which nitrogen has been initially assimilated, i.e. glutamine and glutamic acid, as principal donors, amino nitrogen can be readily mobilised via transamination reactions catalysed by enzymes known as aminotransferases (these enzymes are also referred to as transaminases). It is well known that higher plants contain a complete range of aminotransferases capable of shuttling assimilated nitrogen between appropriate 2-oxo acids.

Transamination reactions

In a transamination reaction, the α-amino group of an amino acid is transferred to a 2-oxo acid:

$$
\underset{\substack{| \\ R_{(A)} \\ \text{Amino acid (A)}}}{H-\overset{\overset{NH_3^+}{|}}{C}-CO_2^-} + \underset{\substack{| \\ R_{(B)} \\ \text{2-Oxo acid (B)}}}{\overset{\overset{O}{\|}}{C}-CO_2^-} \rightleftharpoons \underset{\substack{| \\ R_{(A)} \\ \text{2-Oxo acid (A)}}}{\overset{\overset{O}{\|}}{C}-CO_2^-} + \underset{\substack{| \\ R_{(B)} \\ \text{Amino acid (B)}}}{H-\overset{\overset{NH_3^+}{|}}{C}-CO_2^-}
$$

Aminotransferases contain pyridoxal phosphate, a vitamin B_6 derivative, as prosthetic group. During the transamination reaction pyridoxal phosphate is transiently converted into pyridoxamine phosphate:

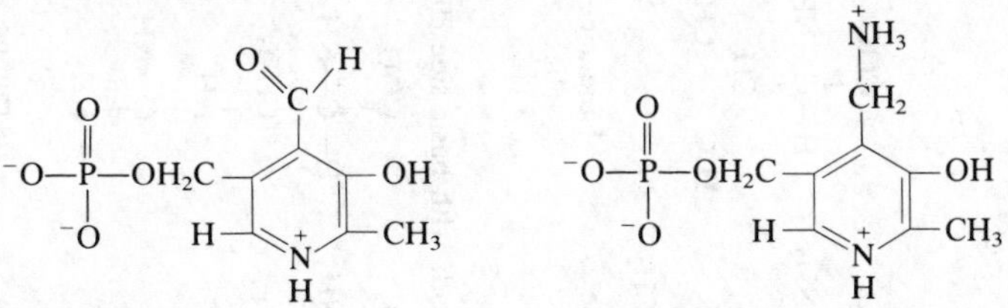

Pyridoxal phosphate   Pyridoxamine phosphate

When substrate is absent, the aldehyde group of pyridoxal phosphate forms a Schiff-base linkage with the side chain (ε) amino group of a specific lysine residue positioned at the active site of the aminotransferase enzyme. However, when an amino acid substrate is present during transamination, the α-amino group of the amino acid displaces the ε-amino group of the lysine residue from the Schiff-base linkage between pyridoxal phosphate and aminotransferase enzyme and forms a Schiff-base between pyridoxal phosphate and the amino acid substrate (Fig. 2.3).

This intermediate remains tightly bound to the aminotransferase enzyme via non-covalent forces. The first half of the transamination reaction illustrated in Fig. 2.3 can be summarised as:

Amino acid (A) + Aminotransferase–Pyridoxal phosphate $\rightleftharpoons$
2-Oxo acid (A) + Aminotransferase–Pyridoxamine phosphate

The second half of the reaction takes place via a reversal of this reaction whereby a second 2-oxo acid (B) reacts with the aminotransferase – pyridoxamine phosphate complex resulting in the formation of an amino acid having the carbon skeleton of the

(i) Amino acids with aliphatic side chains

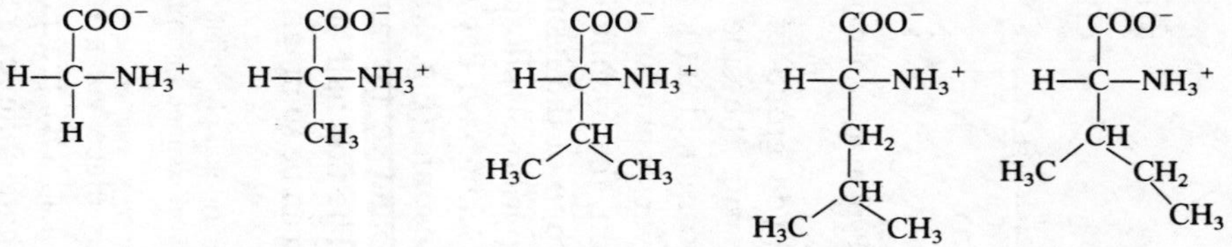

Glycine (Gly)

$$COO^-$$
$$H-C-NH_3^+$$
$$H$$

Alanine (Ala)

$$COO^-$$
$$H-C-NH_3^+$$
$$CH_3$$

Valine (Val)

$$COO^-$$
$$H-C-NH_3^+$$
$$CH$$
$$H_3C \quad CH_3$$

Leucine (Leu)

$$COO^-$$
$$H-C-NH_3^+$$
$$CH_2$$
$$CH$$
$$H_3C \quad CH_3$$

Isoleucine (Ile)

$$COO^-$$
$$H-C-NH_3^+$$
$$CH$$
$$H_3C \quad CH_2$$
$$CH_3$$

(ii) Amino acids with aliphatic hydroxyl side chains

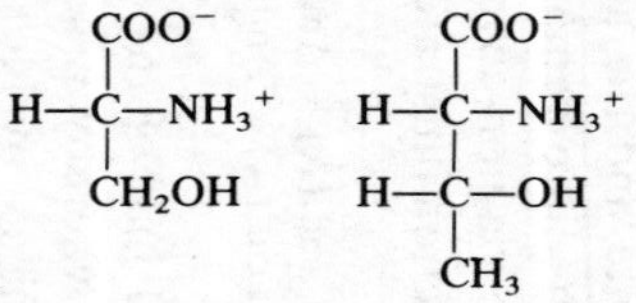

Serine (Ser)

$$COO^-$$
$$H-C-NH_3^+$$
$$CH_2OH$$

Threonine (Thr)

$$COO^-$$
$$H-C-NH_3^+$$
$$H-C-OH$$
$$CH_3$$

(iii) Amino acids with basic side chains

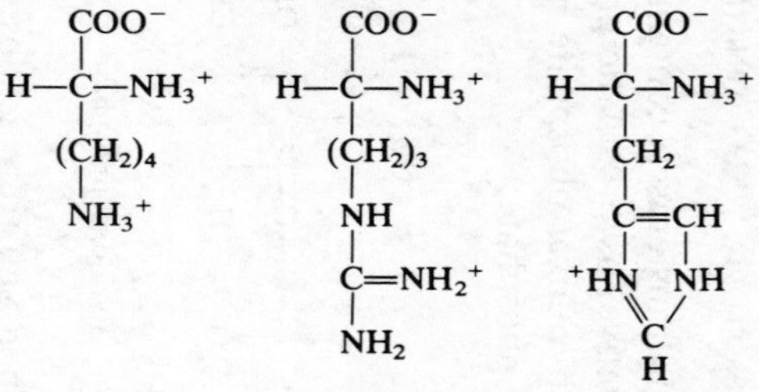

Lysine (Lys)

$$COO^-$$
$$H-C-NH_3^+$$
$$(CH_2)_4$$
$$NH_3^+$$

Arginine (Arg)

$$COO^-$$
$$H-C-NH_3^+$$
$$(CH_2)_3$$
$$NH$$
$$C=NH_2^+$$
$$NH_2$$

Histidine (His)

$$COO^-$$
$$H-C-NH_3^+$$
$$CH_2$$
$$C=CH \quad NH$$
$$^+HN \quad C$$
$$H$$

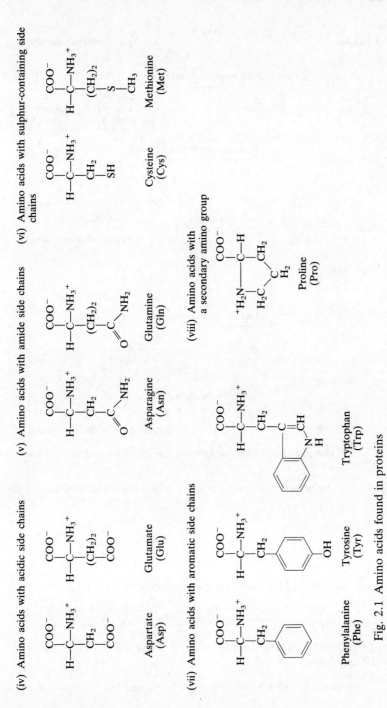

Fig. 2.1 Amino acids found in proteins

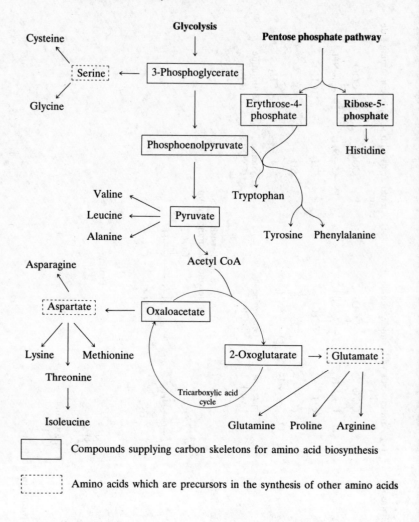

Compounds supplying carbon skeletons for amino acid biosynthesis

Amino acids which are precursors in the synthesis of other amino acids

2-oxo acid (B). This half reaction can be summarised as:

2-Oxo acid (B) + Aminotransferase–Pyridoxamine phosphate $\rightleftharpoons$
Amino acid (B) + Aminotransferase–Pyridoxal phosphate

The overall transamination reaction is thus the sum of these two half reactions:

Amino acid (A) + 2-Oxo acid (B) $\rightleftharpoons$ 2-Oxo acid (A) + Amino acid (B)

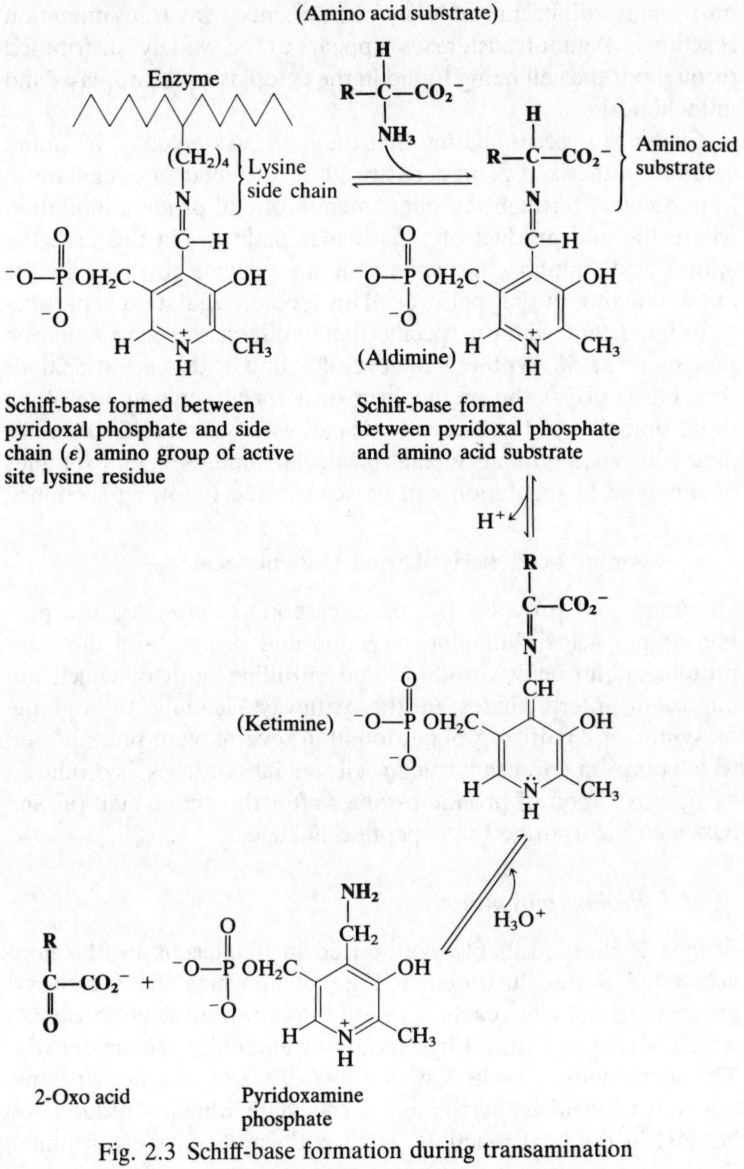

Fig. 2.3 Schiff-base formation during transamination

Plant cells are known to contain many different oxo acids and some of the aminotransferases found in plant tissues have a wide range of substrate specificity, thus once the nitrogen atom has been assimilated into an amino group it can be readily redistributed into many different cellular components via transamination reactions. Aminotransferases appear to be widely distributed throughout the cell being found in the cytoplasm, chloroplasts and mitochondria.

Evidence is accumulating that the pathways involved in amino acid biosynthesis in plant cells are subject to feedback regulation, in particular, through the phenomenon of end product inhibition where the end product of a particular pathway, in this case the amino acid, inhibits the activity of the enzyme involved in the initial reaction of that pathway. This type of regulation is particularly important when a specific metabolite serves as a common precursor for the synthesis of several amino acids since it enables the cell to divert the metabolism of a specific metabolite along those branches of the biosynthetic pathways which fulfil the cell's metabolic requirements at that particular time. Several examples of this type of regulation will be seen in the following sections.

Amino acids derived from glutamic acid

Glutamic acid provides the basic carbon skeleton for the protein amino acids glutamine, arginine and proline and the non-protein amino acids ornithine and citrulline both of which are important intermediates in the pathways leading to arginine biosynthesis. Hydroxyproline, found in several plant proteins and which plays an important role in cell wall biosynthesis, is produced by hydroxylation of proline residues after the amino acid proline has been incorporated into peptide linkage.

Proline biosynthesis

Proline is most probably synthesised in plant cells by the non-acetylated route illustrated in Fig. 2.4 in which the γ-carboxyl group of glutamate reacts with ATP to form an acyl phosphate which is turn is reduced by NADH to glutamic γ-semialdehyde. The semialdehyde cyclises with a loss of water in a non-enzymic reaction to yield Δ'-pyrroline-5-carboxylate which is reduced by NADH in the final reaction of the pathway to produce proline.

Arginine biosynthesis

The biosynthesis of arginine involves intermediates which are acetylated derivatives of glutamate (Fig. 2.4). Blocking of the amino group of glutamate prior to reduction to the semialdehyde prevents the cyclisation step seen in the proline biosynthetic pathway and allows the γ-aldehyde group to be transaminated to an amino group with the consequent production of the non-protein amino acid ornithine after removal of the acetyl blocking group. Most plants have the capability of using either acetyl coenzyme A or *N*-acetyl ornithine as acetyl donors in the reaction leading to the formation of *N*-acetyl glutamate thus allowing the use of acetyl CoA to prime the pathway and then conserving this acetyl group by using *N*-acetyl ornithine, an intermediate of the pathway, as acetyl donor in subsequent acetylation reactions to produce N-acetyl glutamate and ornithine (Fig. 2.4). An unusual feature of the transamination reaction in the pathway leading from *N*-acetyl glutamate to ornithine is that an aldehyde group rather than an oxo group is involved. Glutamate is the preferred amino donor in this reaction. Ornithine now serves as a precursor for arginine biosynthesis utilising the enzymes of the urea cycle (Fig. 2.5). All the enzymes of the cycle have been demonstrated to be present in higher plants. The initial reaction of the urea cycle involves a condensation reaction between ornithine and carbamoyl phosphate to produce another non-protein amino acid, citrulline. Carbamoyl phosphate serves as an important intermediate not only for arginine biosynthesis but also for pyrimidine biosynthesis (see Chapter 4) and is synthesised from glutamine, CO_2 and ATP:

$$Glutamine + 2ATP + HCO_3^- \longrightarrow$$

$$\underset{\text{(Carbamoyl phosphate)}}{H_2N-\overset{\overset{\text{O}}{\|}}{C}-O-\overset{\overset{\text{O}}{\|}}{\underset{\underset{\text{O}^-}{|}}{P}}-O^-} + 2ADP + P_i + Glutamate$$

Only one carbamoyl phosphate synthetase enzyme is present in plants in contrast to mammalian systems and fungi which possess two separate enzymes, one for use in pyrimidine biosynthesis and the other for urea cycle functioning.

Citrulline now condenses with another amino acid, aspartate, in a reaction in which ATP is hydrolysed to AMP and pyrophosphate (Fig. 2.5) and from the product of this reaction, argininosuccinate, the tricarboxylic acid cycle intermediate fumarate is eliminated (the fumarate produced corresponding to the carbon skeleton of aspartate) and arginine is produced. The four nitrogen atoms found in arginine are thus derived from glutamate (2), the amide group of glutamine via carbamoyl phosphate (1) and aspartate (1).

38

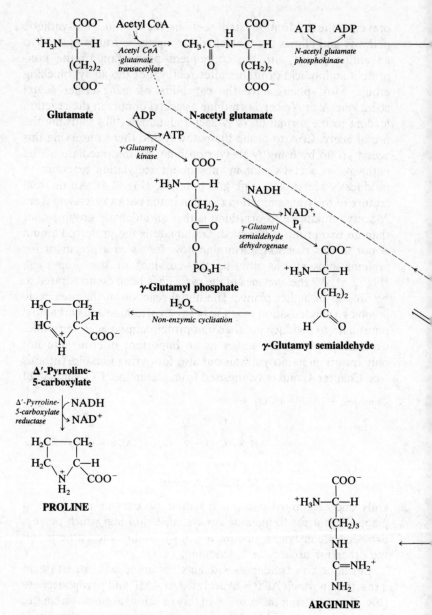

Fig. 2.4 The biosynthesis of proline and ornithine

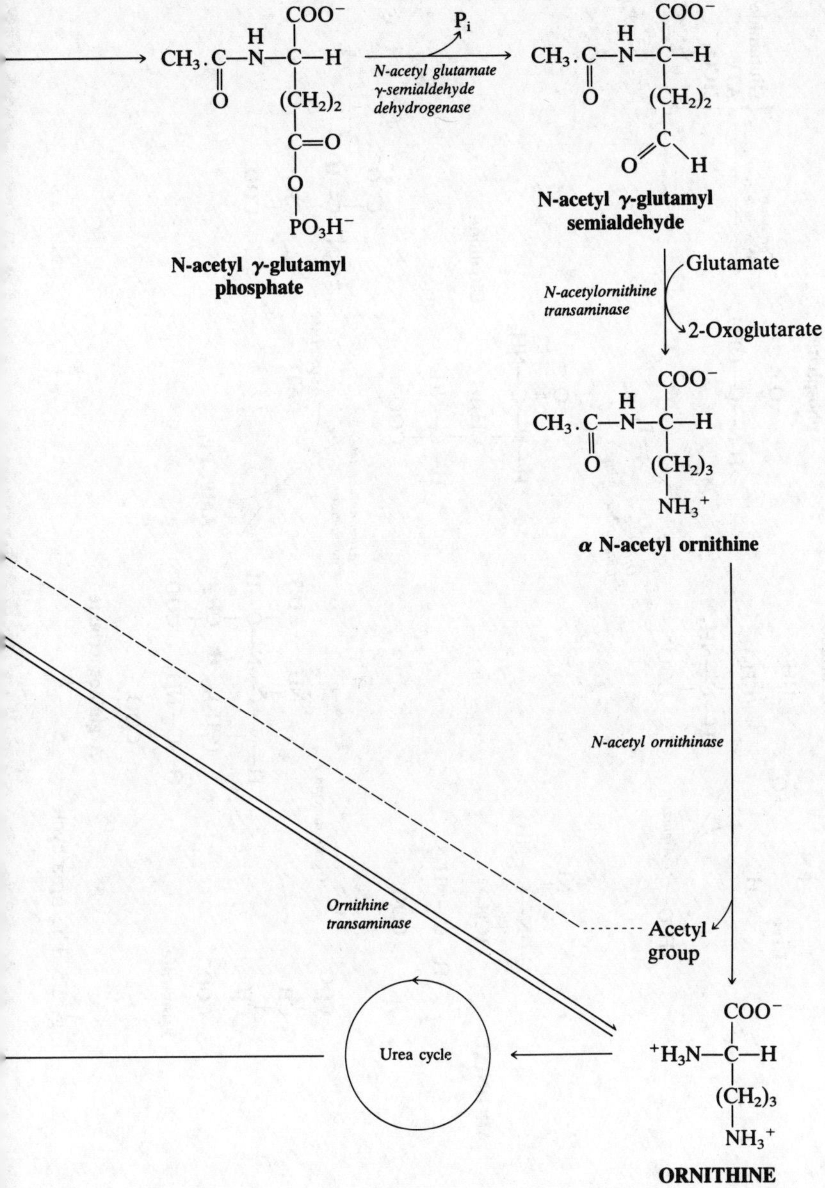
COO⁻
H
CH3.C—N—C—H
O
(CH2)2
C=O
O
PO3H⁻
N-acetyl γ-glutamyl phosphate
Pi
N-acetyl glutamate γ-semialdehyde dehydrogenase
N-acetyl γ-glutamyl semialdehyde
Glutamate
N-acetylornithine transaminase
2-Oxoglutarate
(CH2)3
NH3⁺
α N-acetyl ornithine
N-acetyl ornithinase
Acetyl group
Ornithine transaminase
Urea cycle
⁺H3N—C—H
ORNITHINE

40

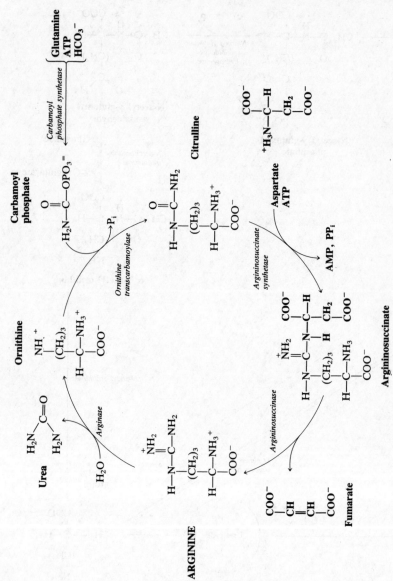

Fig. 2.5 The urea cycle

The first step in arginine biosynthesis, namely the acetylation of glutamate is subject to feedback regulation by ornithine, citrulline and arginine which inhibit the acetylation process. Carbamoyl phosphate synthetase is inhibited by uridylic acid (UMP) and activated by ornithine, thus in the presence of the pyrimidine nucleotide UMP, carbamoyl phosphate synthesis is inhibited until ornithine concentrations in the cell increase to the level at which they can overcome this inhibition by UMP. As UMP also inhibits aspartate transcarbamoylase activity, pyrimidine biosynthesis only recommences when the levels of UMP in the cell fall. Thus these control mechanisms provide a means whereby the cell can channel nitrogen into either arginine formation or pyrimidine biosynthesis.

Amino acids derived from aspartic acid

Aspartate can be synthesised by a transamination reaction utilising glutamate as the amino donor and oxaloacetate, a tricarboxylic acid cycle intermediate, as the accepting 2-oxo acid:

$$
\begin{array}{ccccc}
\text{COO}^- & \text{COO}^- & & \text{COO}^- & \text{COO}^- \\
| & | & & | & | \\
(\text{CH}_2)_2 & \text{CH}_2 & \text{Aspartate} & (\text{CH}_2)_2 & \text{CH}_2 \\
| & | & \text{aminotransferase} & | & | \\
\text{H}-\text{C}-\text{NH}_3^+ + & \text{C}=\text{O} & \rightleftharpoons & \text{C}=\text{O} & + \text{ H}-\text{C}-\text{NH}_3^+ \\
| & | & & | & | \\
\text{COO}^- & \text{COO}^- & & \text{COO}^- & \text{COO}^- \\
\text{Glutamate} & \text{Oxaloacetate} & & \text{2-Oxoglutarate} & \text{Aspartate}
\end{array}
$$

Asparagine biosynthesis

A second nitrogen atom can be introduced into the aspartate molecule by the enzyme asparagine synthetase. This enzyme catalyses the amidation of aspartate using glutamine as amino donor to give the amino acid asparagine:

$$
\begin{array}{l}
\text{COO}^- \\
| \\
\text{CH}_2 \quad + \text{ Glutamine } + \text{ ATP } \longrightarrow \\
| \\
\text{H}-\text{C}-\text{NH}_3^+ \\
| \\
\text{COO}^- \\
\text{Aspartate}
\end{array}
$$

$$
\begin{array}{l}
\text{NH}_2 \\
| \\
\text{C}=\text{O} \\
| \\
\text{CH}_2 \quad + \text{ Glutamate } + \text{ AMP } + \text{ PP}_i \\
| \\
\text{H}-\text{C}-\text{NH}_3^+ \\
| \\
\text{COO}^- \\
\text{Asparagine}
\end{array}
$$

Asparagine synthetases have been isolated from several plant tissues and the enzyme has been shown to have a low K_m for glutamine but a high K_m for ammonia. Asparagine accumulates to quite a high concentration in many plant cells.

Asparagine synthesis by a second route may serve as a method of cyanide detoxification. This route requires CN^- and the enzymes β-cyanoalanine synthetase and β-cyanoalanine hydrolase which catalyse the reactions:

$$\text{Cysteine} + CN^- \rightarrow \beta\text{-Cyanoalanine} + H_2S$$

$$\beta\text{-Cyanoalanine} + H_2O \rightarrow \text{Asparagine}$$

Plant tissues, e.g. lupin seedlings, which possess this capability for cyanide detoxification also synthesise asparagine by the route using aspartate and asparagine synthetase.

Lysine, threonine, isoleucine and leucine biosynthesis

In addition to being one of the precursors of pyrimidine biosynthesis (see Chapter 4), aspartate is also the initial precursor molecule in the synthesis of the amino acids lysine, threonine, isoleucine and methionine. Aspartate is initially converted into the semialdehyde by the enzymes aspartate kinase and aspartate β-semialdehyde dehydrogenase in reactions analogous to those involved in the glutamate to glutamic semialdehyde conversion. At this point the methionine and threonine biosynthetic pathway branches off from the diaminopimelate pathway (so-called after one of its key intermediates) responsible for lysine biosynthesis in higher plants (Figs. 2.6 and 2.7).

Methionine and threonine then share a common pathway via homoserine to O-phosphohomoserine at which point the two biosynthetic pathways diverge. Homoserine is an amino acid which is rarely detected in the soluble amino acid pool but which is the predominant soluble amino acid in germinating peas. Threonine is formed from O-phosphohomoserine by a reaction involving the elimination of inorganic phosphate and migration of the hydroxyl group from the γ- to the β-carbon atom of the amino acid side chain (Fig. 2.6). Threonine in turn serves as a precursor for the amino acid isoleucine whose biosynthesis is discussed later.

The sulphur atom of the amino acid methionine is introduced into the biosynthetic pathway via cysteine which in plant tissues reacts with O-phosphohomoserine in the presence of the enzyme cystathionine γ-synthase to produce the thioether cystathionine, a central intermediate in trans-sulphuration (Fig. 2.6). The enzyme cystathionine β-lyase, which is found widely distributed in plant tissues, then catalyses a β-elimination reaction to produce homocysteine, pyruvate and ammonia from cystathionine. The final step in methionine biosynthesis is the introduction of a methyl group, the methyl donor being most probably a folic acid derivative, these derivatives being highly versatile carriers of activated one-carbon units for use in biosynthetic reactions. Methionine is an amino acid which plays an important metabolic role not only as a component of proteins but also as a methyl donor (in the form of S-adenosyl methionine) in many cellular reactions and has an important role to play in the initiation of protein synthesis (see Chapter 5).

Lysine biosynthesis proceeds from β-aspartyl semialdehyde by operation of the diaminopimelate pathway (Fig. 2.7). This series of reactions involves an initial condensation of pyruvate with the semialdehyde to produce dihydropicolinate which undergoes successive reduction, succinylation, transamination and deacylation reactions to produce diaminopimelate. A final decarboxylation reaction produces lysine. The second nitrogen atom of lysine is introduced during the transamination reaction. The first enzyme unique to lysine biosynthesis, dihydropicolinate synthase, and the last enzyme of the pathway, diaminopimelate decarboxylase have both been shown to be localised in the chloroplasts in green leaves but the presence of the intervening enzymes of the pathway in plant tissue has yet to be demonstrated. It is probable that the intermediate enzymes of the pathway are also localised solely in the chloroplasts in green leaf tissue although this still remains to be confirmed.

Several control points exist in this branched biosynthetic pathway. Aspartate kinase, the first enzyme common to the biosynthesis of lysine, methionine, threonine and isoleucine exists as isoenzymes sensitive to inhibition by lysine or threonine which can thus inhibit their own synthesis by controlling the conversion of aspartate through to β-aspartyl phosphate. In some plant tissues the presence of methionine or S-adenosyl methionine may enhance the inhibitory effects of lysine on aspartate kinase. However, no overall general pattern of control is apparent when different plant tissues are compared. Lysine can also control its own synthesis through the high sensitivity of dihydropicolinate synthase, the first enzyme of the branch of the pathway leading to

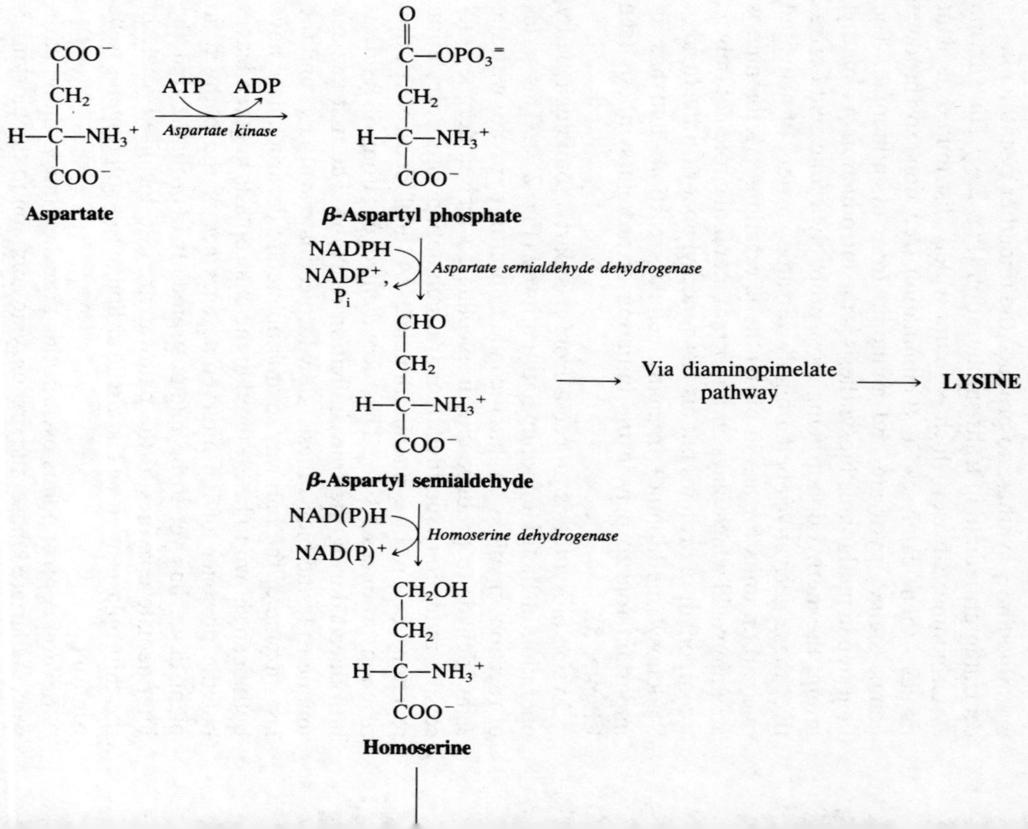

Aspartate

$$COO^-$$
$$CH_2$$
$$H-C-NH_3^+$$
$$COO^-$$

ATP → ADP, *Aspartate kinase*

β-Aspartyl phosphate

$$O=C-OPO_3^=$$
$$CH_2$$
$$H-C-NH_3^+$$
$$COO^-$$

NADPH, NADP$^+$, P$_i$, *Aspartate semialdehyde dehydrogenase*

β-Aspartyl semialdehyde

$$CHO$$
$$CH_2$$
$$H-C-NH_3^+$$
$$COO^-$$

Via diaminopimelate pathway → **LYSINE**

NAD(P)H, NAD(P)$^+$, *Homoserine dehydrogenase*

Homoserine

$$CH_2OH$$
$$CH_2$$
$$H-C-NH_3^+$$
$$COO^-$$

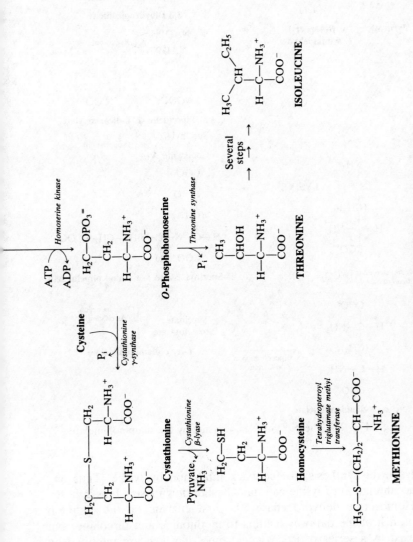

Fig. 2.6 The synthesis of threonine and methionine from aspartate

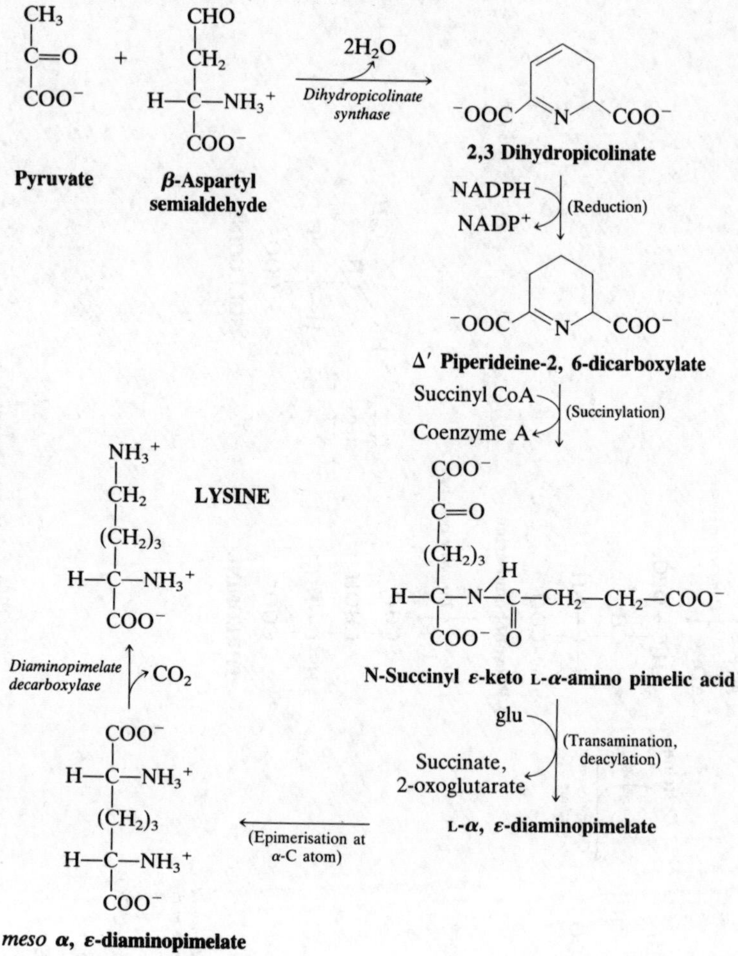

Fig. 2.7 The biosynthesis of lysine by the diaminopimelate
pathway

lysine biosynthesis, to feedback inhibition by lysine. Thus any accumulation of lysine will quickly shut down lysine biosynthesis. Homoserine dehydrogenase, the first enzyme on the common branch of the pathway leading to methionine and threonine, can exist as a series of isoenzymes some of which are inhibited by threonine and control of the activity of this enzyme by threonine may thus regulate the flow of β-aspartyl semialdehyde along this branch of the pathway. The activity of threonine synthase may be under the control of the intracellular levels of cysteine and

S-adenosyl methionine. High levels of cysteine (which is involved in methionine biosynthesis) inhibit threonine synthase while the presence of S-adenosyl methionine has a stimulatory effect on the enzyme. These controls ensure that the conversion of *O*-phosphohomoserine to threonine is very slow when a situation of high cysteine levels and low S-adenosyl methionine levels exist in the cell. Inhibition of threonine synthase in this manner should allow the channelling of *O*-phosphohomoserine into methionine biosynthesis and subsequently S-adenosyl methionine biosynthesis. Thus when a critical cellular level of S-adenosyl methionine is reached, the inhibition of threonine synthase will be released and threonine synthesis allowed to continue.

Amino acids with aliphatic side chains

The pathways leading to the synthesis of the amino acids isoleucine, valine and leucine are outlined in Fig. 2.8.

Threonine acts as a precursor for isoleucine biosynthesis while pyruvate, a glycolytic intermediate, serves as the precursor molecule for valine, leucine and isoleucine biosynthesis. The predominant feature of this biosynthetic pathway is the catalysis of analogous steps of isoleucine and valine biosynthesis by the same group of enzymes. These enzymes catalyse an initial condensation reaction, a reduction step involving a simultaneous intramolecular alkyl migration, a dehydration step and a final transamination reaction to produce either isoleucine or valine. Leucine biosynthesis shares the same pathway as valine as far as the formation of 2-oxoisovalerate (Fig. 2.8) when this intermediate instead of being involved in a transamination reaction to produce valine condenses with acetyl coenzyme A to produce 2-isopropylmalate. An isomerisation step followed by an oxidative decarboxylation reaction converts 2-isopropylmalate into 2-oxoisocaproate which participates in a transamination reaction to produce leucine. In each of these syntheses a 2-oxo acid having the appropriate carbon skeleton is first synthesised prior to the introduction of the amino group via a transamination reaction in the final step.

Acetolactate synthase can be considered to be the first enzyme in a triple-branched pathway and in contrast to aspartokinase the control of this enzyme does not appear to differ significantly between species. Both leucine and valine inhibit acetolactate synthase activity while leucine also exhibits feedback inhibition on isopropylmalate synthase, the first enzyme in the branch of the

48

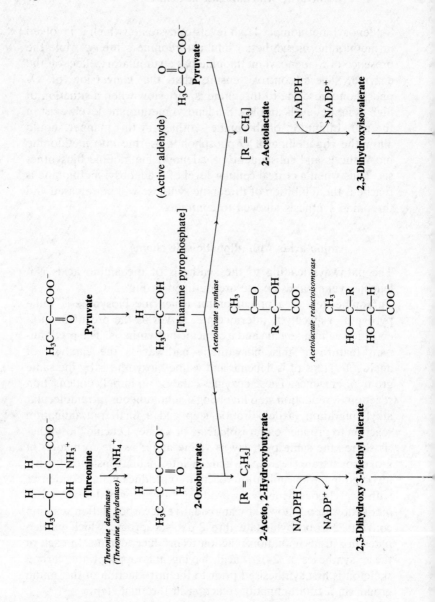

49

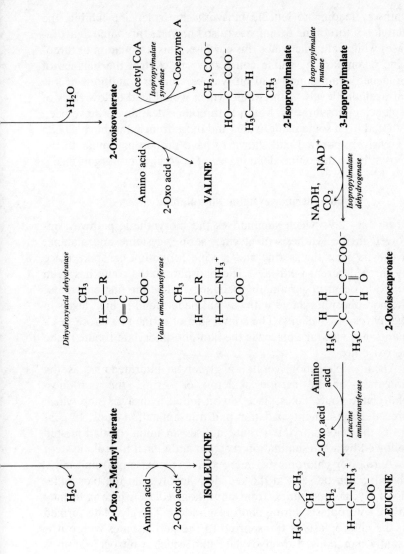

Fig. 2.8 Biosynthesis of amino acids with aliphatic side chains

pathway leading to leucine biosynthesis. Isoleucine inhibits the activity of threonine deaminase (also known as threonine dehydratase) while valine activates this enzyme. This stimulation of threonine deaminase by valine could serve to control the balance of branched chain amino acid biosyntheses under conditions of low intracellular levels of threonine which would lead to low rates of isoleucine biosynthesis. Multiple threonine deaminases have been isolated from some plant tissues and those from spinach have been extensively purified and shown to have properties similar to the biosynthetic threonine deaminases of several micro-organisms.

Glycine, serine, cysteine and alanine

From Fig. 2.9, which summarises the biosynthetic pathways involved in the synthesis of glycine, serine, cysteine and alanine, it can be seen that serine and glycine formation can take place by three different pathways. The phosphorylated route has been found to occur in germinating seedlings, while the first enzyme of the non-phosphorylated pathway has been found in high levels in the leaves of C_4 plants. The conversion of serine to sucrose in C_3 plants may also take place via the non-phosphorylated route in the peroxisome.

Using 3-phosphoglycerate, a glycolytic intermediate, as the precursor of the carbon skeleton of serine, the non-phosphorylated route takes place via an initial hydrolysis step which precedes subsequent oxidation and transamination steps. In contrast, the phosphorylated route utilises an initial oxidation step followed by a transamination reaction and a final hydrolytic step.

A role for chloroplasts, microbodies (peroxisomes) and mitochondria can be seen in the reactions involved in the biosynthesis of glycine and serine from phosphoglycolate which is formed in the chloroplasts during photorespiration. The glycolate formed in the chloroplasts is transported to the peroxisomes where it is rapidly metabolised to glyoxylate into which a nitrogen atom is introduced in a largely unidirectional transamination reaction to produce glycine. Glycine can be a major precursor of serine in a mitochondrial reaction in which two molecules of glycine give rise to one molecule of serine with the elimination of carbon dioxide and ammonia. This complex reaction involves the decarboxylation and deamination of one of the glycine molecules and the transfer of the remaining one-carbon hydroxy methyl residue as a folic acid derivative (5,10-methylenetetrahydrofolate) to the second glycine molecule. The reaction is catalysed by the enzyme serine hydroxymethyl transferase. This synthesis of serine from glycine in the mitochondria can also be coupled to ATP generation. The ammonia released in this reaction would be reassimilated by either the mitochondrial glutamate dehydrogenase or cytosolic or plastid glutamine synthetase.

Control mechanisms have currently been demonstrated to operate on only one of the many pathways for serine and glycine biosynthesis. 3-Phosphoglycerate dehydrogenase, the first enzyme of the phosphorylated pathway for serine biosynthesis, has been shown to be subject to end product inhibition by serine in peas.

The formation of cysteine from serine requires the substitution of a sulphur atom for the serine side chain oxygen atom. The initial activation of serine to give O-acetyl serine is followed by a reaction which incorporates sulphide into this molecule combined with the simultaneous release of acetate. The nature of the sulphydryl donor in higher plants is uncertain, but since free sulphide is a potent inhibitor of many enzymes involved in aerobic respiration, it is probably 'bound' sulphide which serves as donor. The acetylated derivative of serine is much more reactive in the sulphydration reaction than is serine alone.

The biosynthesis of alanine in plant cells utilises pyruvate as principal precursor (cf. valine and leucine biosynthesis). Pyruvate acts as the amino acceptor in a transamination reaction catalysed by alanine aminotransferase, an enzyme which in plant tissues can use a wide variety of amino acids as the amino donor. Alanine biosynthesis is thus regulated, most probably by the availability of pyruvate.

Aromatic amino acids

Tyrosine, phenylalanine and tryptophan

The biosynthetic pathway responsible for the synthesis of the aromatic amino acids tyrosine, phenylalanine and tryptophan in bacteria, the shikimic acid pathway so-called after one of its key intermediate compounds, is outlined in Fig. 2.10. There is good suggestive evidence that this pathway is also operative in plant cells. The first part of the pathway is common to all three amino acids and involves the synthesis of chorismate from D-erythrose-4-phosphate and two molecules of phosphoenolpyruvate. The pathways for tryptophan biosynthesis and phenylalanine/tyrosine biosynthesis diverge at this point.

52

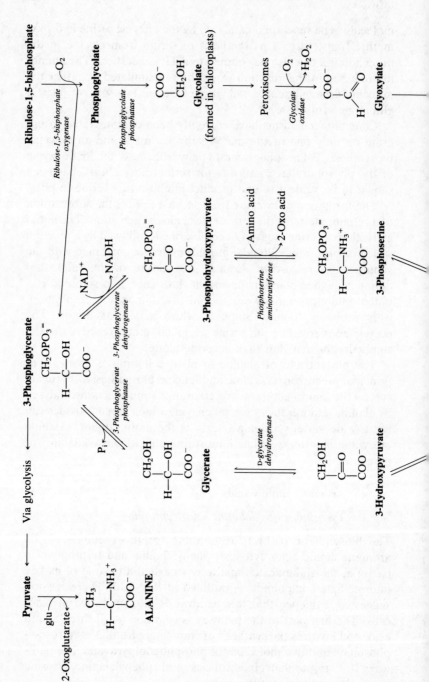

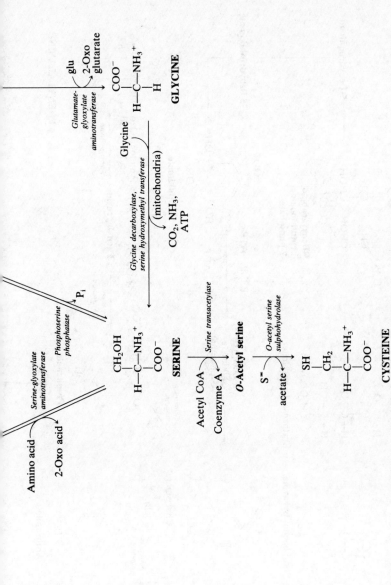

Fig. 2.9 Biosynthesis of glycine, serine, cysteine and alanine

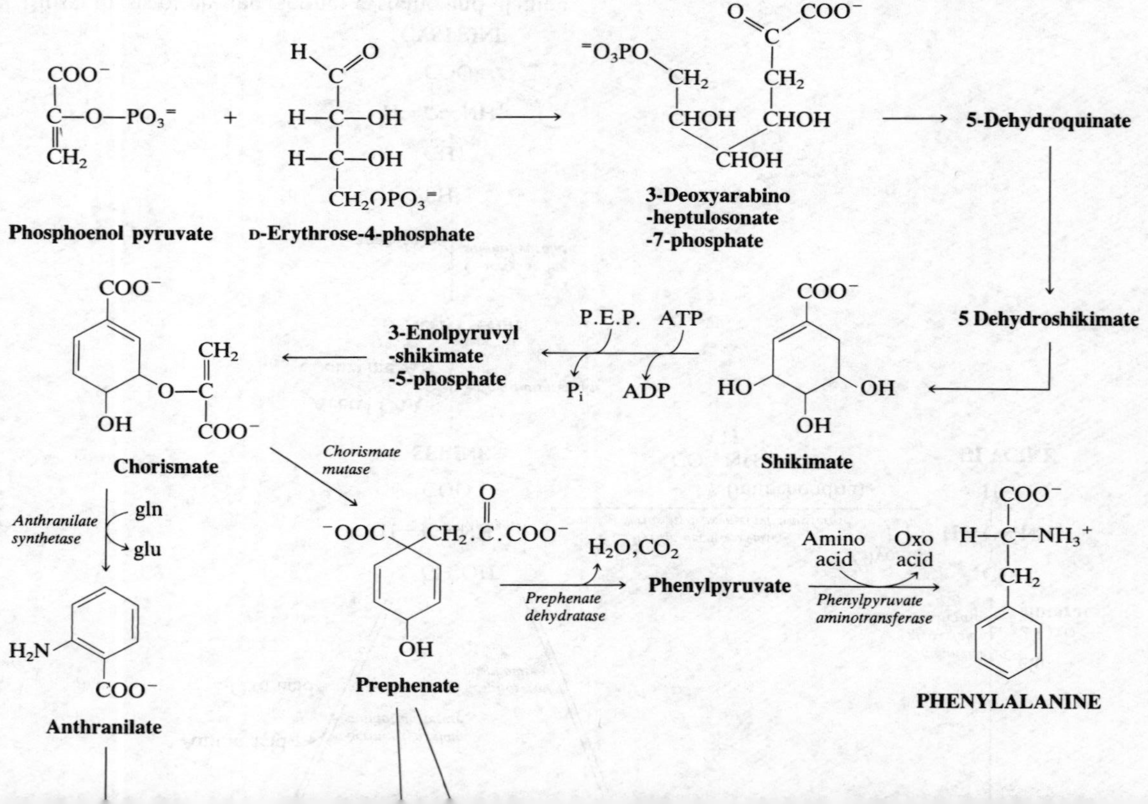

54

Phosphoenol pyruvate

D-Erythrose-4-phosphate

3-Deoxyarabino-heptulosonate-7-phosphate

5-Dehydroquinate

5 Dehydroshikimate

Shikimate

P.E.P. ATP

ADP

P_i

3-Enolpyruvyl-shikimate-5-phosphate

Chorismate

Chorismate mutase

Prephenate

Prephenate dehydratase

H_2O, CO_2

Phenylpyruvate

Amino acid

Oxo acid

Phenylpyruvate aminotransferase

PHENYLALANINE

Anthranilate synthetase

gln

glu

Anthranilate

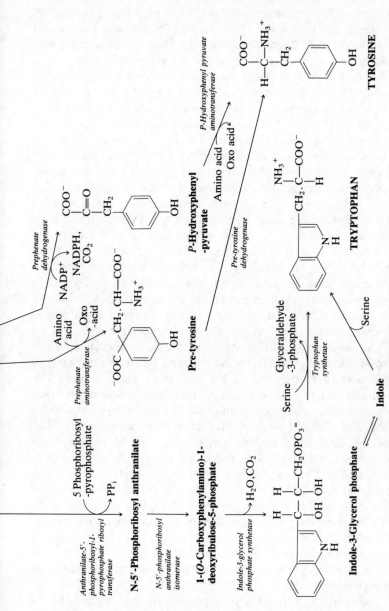

Fig. 2.10 Biosynthesis of phenylalanine, tyrosine and tryptophan

The enzymes responsible for the conversion of chorismate to tryptophan have been demonstrated to be present in plant tissues and the initial reaction of this branch of the pathway is catalysed by anthranilate synthetase which incorporates an amino group into chorismate to produce anthranilate. Glutamine is used as the amino donor. Anthranilate then condenses with 5'-phosphoribosyl pyrophosphate to produce an amino glycoside of anthranilic acid which undergoes an isomerisation reaction followed by a reaction involving dehydration and decarboxylation steps to produce indole-3-glycerol phosphate. Finally, tryptophan synthetase catalyses the replacement of the glycerol phosphate side chain of indole-3-glycerol phosphate by the carbon skeleton and amino group of serine to produce tryptophan. Purified tryptophan synthetase from peas has been shown to comprise two components, A and B. Both components are required for the overall synthetic reaction: component A catalyses the formation of indole and glyceraldehyde-3-phosphate from indole-3-glycerol phosphate; and component B catalyses the direct condensation between indole and serine to form tryptophan (see Fig. 2.10). Control of tryptophan biosynthesis is mediated by feedback inhibition of the first enzyme of this branch of the pathway, namely the glutamine-dependent anthranilate synthetase, by tryptophan.

The enzymes responsible for the synthesis of phenylalanine and tyrosine from chorismate by the reactions illustrated in Fig. 2.10 have been demonstrated to be present in plant tissues. The production of the common intermediate prephenate is catalysed by the enzyme chorismate mutase which exists as two isoenzymes (CM_1 and CM_2) in some plants and as three isoenzymes (CM_1, CM_2 and CM_3) in others. Isoenzymes CM_1 and CM_3 are subject to inhibition by phenylalanine and tyrosine and activation by tryptophan while CM_2 exhibits none of these properties. It has been suggested that CM_2 may be responsible for the production of phenylalanine and tyrosine for channelling into metabolic pathways responsible for the production of secondary metabolites. Phenylalanine is synthesised from prephenate via a reaction involving dehydration and decarboxylation steps to produce the oxo acid phenylpyruvate which can be readily transaminated to phenylalanine.

The synthesis of tyrosine from prephenate may follow either of the two routes illustrated in Fig. 2.10. An initial dehydrogenation–decarboxylation reaction to yield *p*-hydroxy phenylpyruvate, an oxo acid, can be followed by a transamination reaction to

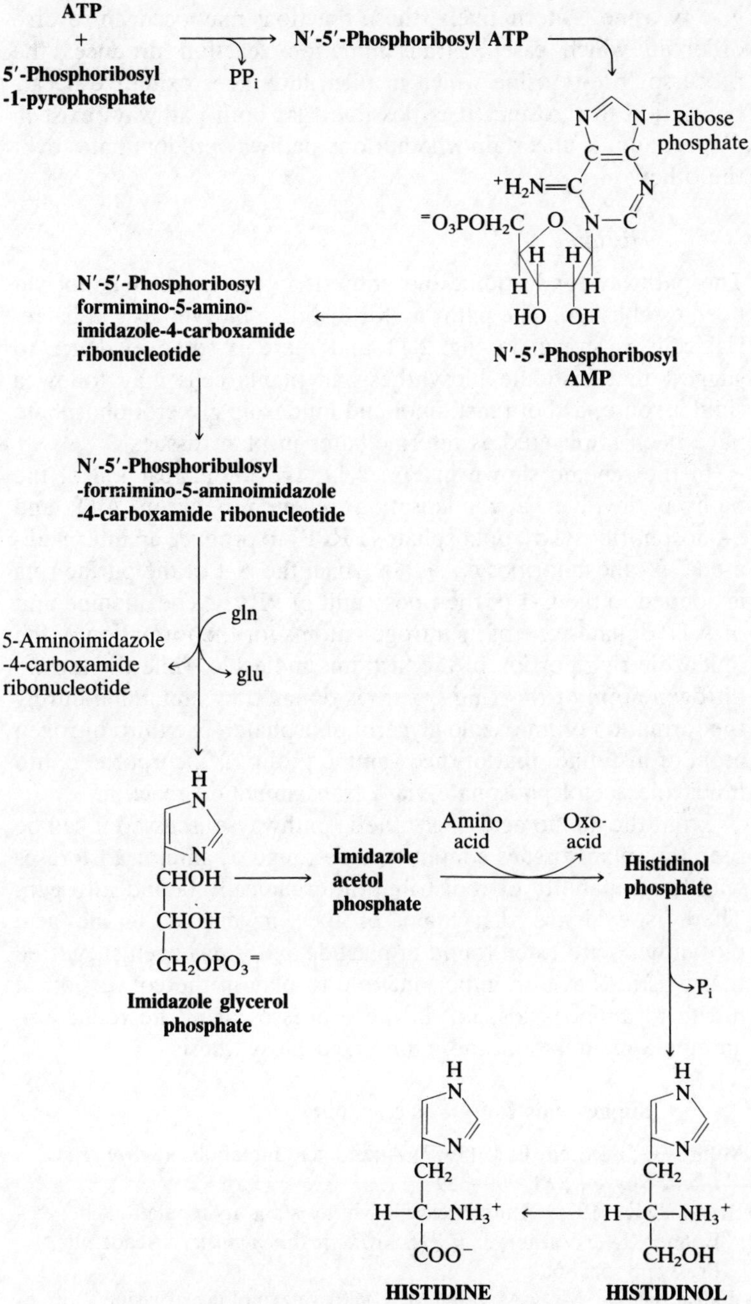

Fig. 2.11 Biosynthesis of histidine

give tyrosine. Alternatively, these reactions may occur in reverse order in which case a transamination reaction produces the precursor pre-tyrosine which in turn undergoes oxidative decarboxylation to tyrosine. It is possible that both pathways exist in plants and it is uncertain whether one pathway predominates over the other.

Histidine

The pathway for histidine biosynthesis in plant cells has not yet been established. The pathway for histidine biosynthesis in bacterial cells is shown in Fig. 2.11 and there is some evidence to suggest that histidine biosynthesis in plants cells may follow a similar route, as both histidinol and imidazole glycerol phosphate have been implicated as intermediates in plant tissues.

In the scheme shown in Fig. 2.11, the initial reaction of the pathway involves a condensation reaction between ATP and 5-phosphoribosyl pyrophosphate (PRPP) to produce an intermediate N'-5'-phosphoribosyl ATP in which the N-1 of the purine ring is bonded to the C-1 of the ribose unit of PRPP. The adenine unit of ATP donates one of its nitrogen atoms for the formation of the imidazole ring portion of the histidine molecule while the second nitrogen atom of this ring system is donated by glutamine during the formation of imidazole glycerol phosphate. The third nitrogen atom of histidine, that of the α-amino group, is incorporated into imidazole acetol phosphate via a transamination reaction.

From the amino acid biosynthetic pathways discussed it can be seen that plant tissues contain a wide range of aminotransferases with the capability of redistributing organically bound nitrogen. There is evidence that many of the enzymes of amino acid biosynthesis are to be found in plastids and it has been suggested that plastids have an important role to play in the conversion of nitrite to amino acids, i.e. in the processes of nitrite reduction, ammonia assimilation and amino acid biosynthesis.

Suggestions for further reading

Miflin, B. J. & Lea, P. J. (1977) Amino acid metabolism, *Ann. Rev. Plant. Physiol.*, **28**, 299–329

Bryan, J. K. (1976) Amino acid biosynthesis and its regulation, in Bonner, J. & Varner, J. E., (eds) *Plant Biochemistry*. Academic Press, pp. 525–60

Wallsgrove, R. M. & Mazelis, M. (1980) Enzymology of lysine biosynthesis in higher plants, *FEBS Letts.*, **116**, 189–92

3

Nitrogen redistribution in cells

The 20 amino acids which serve as substrates for the synthesis of proteins are also precursors of a variety of other large and small biomolecules. The nitrogen contained in these protein amino acids can be redistributed via the metabolic pathways of the plant cell into non-protein amino acids such as canavanine, homoserine and ornithine and into other compounds which include γ-amino butyric acid, primary amines, diamines, aldehydes and hydroxylated amino acids which in turn may serve as intermediates in many of the anabolic and catabolic processes in the cell.

Many of the cellular reactions involving amino acids and amines are catalysed by enzymes which require pyridoxal phosphate as coenzyme. Pyridoxal phosphate enzymes labilise one of three bonds at the α-carbon atom of an amino acid substrate (Fig. 3.1) giving rise to transamination, decarboxylation or aldol cleavage reactions. Pyridoxal-phosphate-containing enzymes such as tryptophan synthetase and cystathionase can also catalyse reactions

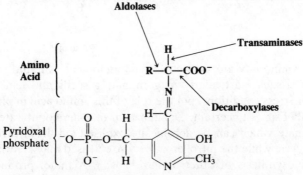

Fig. 3.1 Bonds labilised by pyridoxal phosphate enzymes

at the β- and γ-carbon atoms respectively, and the role of the pyridoxal phosphate coenzyme in Schiff-base formation during transamination reactions has been discussed in Chapter 2. Other reactions occurring at the α-carbon atom of amino acids include deamination and racemisation reactions.

Transamination reactions

Transamination reactions play a central role in nitrogen metabolism and are discussed in Chapters 2 and 6. These reactions are usually readily reversible reactions allowing extensive interconversion of amino acids as seen in the variations in the composition of the amino acid pool during seed maturation and germination.

Decarboxylation reactions

The decarboxylation of amino acids is an essentially irreversible process and often occurs as the final step in the synthesis of amino compounds such as γ-amino butyrate and β-alanine:

$$
\begin{array}{c}
\text{COO}^- \\
| \\
\text{CH}_2 \\
| \\
\text{CH}_2 \\
| \\
\text{CHNH}_3{}^+ \\
| \\
\text{COO}^-
\end{array}
\quad \xrightarrow[\text{CO}_2]{\alpha\text{-Decarboxylation}} \quad
\begin{array}{c}
\text{COO}^- \\
| \\
\text{CH}_2 \\
| \\
\text{CH}_2 \\
| \\
\text{CH}_2 \\
| \\
\text{NH}_3{}^+
\end{array}
$$

Glutamate $\qquad\qquad$ γ-Aminobutyrate

$$
\begin{array}{c}
\text{COO}^- \\
| \\
\text{CH}_2 \\
| \\
\text{CHNH}_3{}^+ \\
| \\
\text{COO}^-
\end{array}
\quad \xrightarrow[\text{CO}_2]{\alpha\text{-Decarboxylation}} \quad
\begin{array}{c}
\text{COO}^- \\
| \\
\text{CH}_2 \\
| \\
\text{CH}_2 \\
| \\
\text{NH}_3{}^+
\end{array}
$$

β-Alanine

γ-Aminobutyrate, a non-protein amino acid, is found in many plant tissues, particularly in germinating seeds and leaf tissues under stress conditions, but the role of this amino acid in plant cell metabolism is uncertain. Deamination of γ-aminobutyrate yields succinate which can be aerobically oxidised via the tricarboxylic acid cycle while the nitrogen moiety so released may be utilised in the biosynthesis of other amino acids. β-Alanine, produced by α-decarboxylation of aspartate, can sometimes be found in the soluble non-protein amino acid fraction of plant cell extracts and has an important metabolic role to play since it forms a part of the coenzyme A molecule.

There are two general mechanisms whereby amines are synthesised in the cell. One reaction involves the α-decarboxylation of an amino acid to produce an amine and can be represented by the general equation:

$$R-\underset{\underset{NH_3^+}{|}}{\overset{\overset{H}{|}}{C}}-COO^- \xrightarrow{\quad CO_2 \quad} R-\underset{\underset{NH_2}{|}}{CH_2}$$

Amino acid   Amine

The type of amine produced depends upon the nature of the side chain (R), an aliphatic primary amine being produced if R is aliphatic in nature, an aromatic amine arising if R is aromatic and a diamine being produced if R is an aliphatic side chain containing an amino group. Examples of the synthesis of these different types of amine are illustrated in Fig. 3.2

The decarboxylation of serine produces ethanolamine which plays an important role in phospholipid biosynthesis along with its methylated derivative choline. The hydroxylation of tyrosine, an aromatic amino acid, produces 3,4-dihydroxyphenylalanine (dopa) which in turn can be decarboxylated to produce the aromatic amine, dopamine, an intermediate in alkaloid biosynthesis while the decarboxylation of the basic amino acid, arginine, results in the production of the diamine, agmatine. Agmatine in turn can be further metabolised via the loss of a carbamoyl moiety to another diamine, putrescine ($NH_2 . (CH_2)_4 . NH_2$).

The second type of reaction which results in amine production involves a transamination reaction between an amino acid and an appropriate aldehyde:

$$R-\underset{\diagdown O}{\overset{\diagup H}{C}} + R'-\underset{\underset{NH_3^+}{|}}{\overset{\overset{H}{|}}{C}}-COO^- \longrightarrow R-CH_2NH_2 + R'-\underset{\underset{O}{\parallel}}{C}-COO^-$$

Aldehyde   Amino acid   Amine   2-Oxo acid

Many of the reactions involved in the production of amines in plant cells have yet to be established but amines usually occur in low concentrations except in flowers where aliphatic primary amines can be found in relatively high concentrations and may

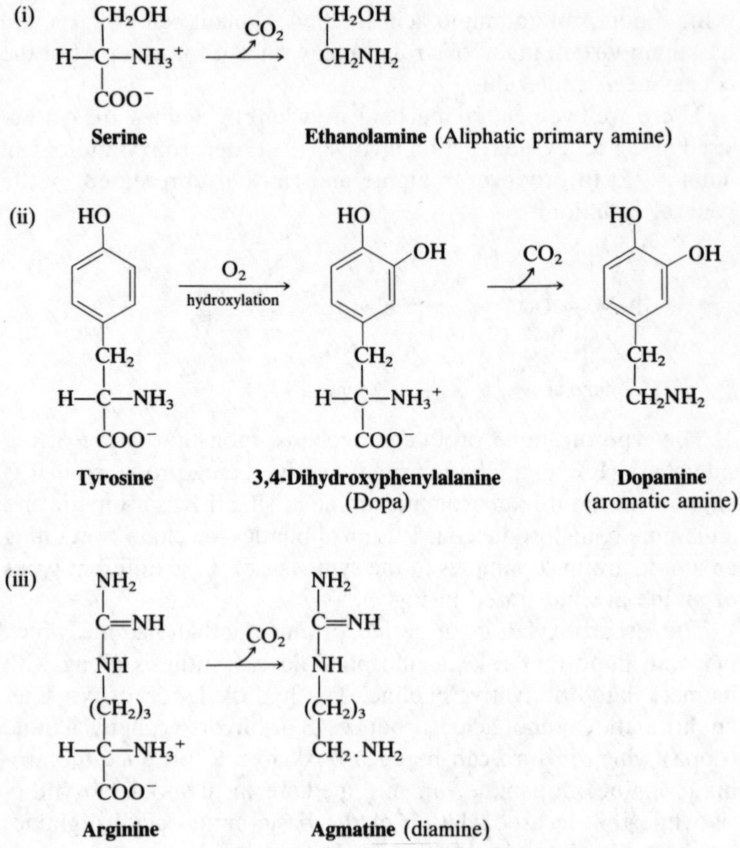

(i)

$$\underset{\text{Serine}}{\overset{\displaystyle CH_2OH}{H-\underset{\displaystyle COO^-}{\overset{\displaystyle |}{\underset{|}{C}}}-NH_3{}^+}} \quad \xrightarrow{\;\;CO_2\;\;} \quad \underset{\text{Ethanolamine (Aliphatic primary amine)}}{\overset{\displaystyle CH_2OH}{\underset{\displaystyle CH_2NH_2}{|}}}$$

(ii)

$$\underset{\text{Tyrosine}}{\text{Tyrosine}} \quad \xrightarrow[\text{hydroxylation}]{O_2} \quad \underset{\substack{\text{3,4-Dihydroxyphenylalanine}\\ \text{(Dopa)}}}{\text{Dopa}} \quad \xrightarrow{\;\;CO_2\;\;} \quad \underset{\substack{\text{Dopamine}\\ \text{(aromatic amine)}}}{\text{Dopamine}}$$

(iii)

$$\underset{\text{Arginine}}{\overset{\displaystyle NH_2}{\underset{\displaystyle COO^-}{\overset{|}{\underset{|}{C}}=NH \atop NH \atop (CH_2)_3 \atop H-C-NH_3{}^+}}} \quad \xrightarrow{\;\;CO_2\;\;} \quad \underset{\text{Agmatine (diamine)}}{\overset{\displaystyle NH_2}{\underset{\displaystyle CH_2.NH_2}{\overset{|}{\underset{|}{C}}=NH \atop NH \atop (CH_2)_3}}}$$

Fig. 3.2 Examples of mono- and diamine biosynthesis from amino acid precursors

play a role in attracting insects, thus assisting in the pollination process. Decarboxylation of the aromatic amino acids phenylalanine and tyrosine yields phenylethylamine and tyramine respectively, these phenolic amines being important metabolites which can be used as precursors in alkaloid biosynthesis.

Deamination reactions

Oxidative

Amines found in plant cells can be oxidatively deaminated by enzymes known as monoamine oxidases or diamine oxidases to yield aldehydes. This type of reaction can be represented by the general equation:

$$R—CH_2.NH_2 + O_2 + H_2O \rightleftharpoons R—C\!\!\begin{array}{c}H\\\\O\end{array} + NH_3 + H_2O_2$$

$$\downarrow \text{\textit{Catalase}}$$

$$H_2O + \tfrac{1}{2}O_2$$

Amine   Aldehyde

The relative importance of these enzymes in amino group metabolism in plant cells is, however, uncertain.

Non-oxidative

In plant cells there are enzymes known collectively as ammonia lyases whose function is to deaminate amino compounds in a non-oxidative fashion. These enzymes are known to play an important role in nitrogen metabolism in plant tissues, and the best characterised enzyme of this type is phenylalanine ammonia lyase (PAL) which catalyses the non-oxidative deamination of both phenylalanine and tyrosine:

Phenylalanine   *trans* cinnamate

Tyrosine   *trans p*-Coumarate

The products of these reactions, *trans* cinnamic acid and *trans* *p*-coumaric acid are important intermediates in the biosynthesis of many secondary metabolites in plants, many of which are indicated in Fig. 3.3. Among the plant products illustrated in this figure are many which give the characteristic fragrance to spices and to certain plants. Also included are the anthocyanidins which derive from cinnamoyl CoA and exist in plants principally as glycosides known as anthocyanins. The colour of these particular compounds depends upon the degree of hydroxylation, the presence or absence of methylation and glycosylation, and pH. Pelargonidin found in the red geranium *Pelargonium* and cyanidin of the blue cornflower *Centaurea cyanus* are examples of anthocyanins of different colour. Yellow flavanol pigments are also derived from cinnamoyl CoA, an example of these pigments being phlorhizin which is found in the root bark of pears, apples and other plants of the rose family.

Pelargonidin   Phlorhizin

The precursors for the biosynthesis of lignins are di- and tri-hydroxy methylated derivatives, synthesis proceeding via an oxidative coupling of coniferyl alcohol and related monomers. The polymerisation process appears to involve a peroxidase enzyme. Lignin, in turn, can be oxidatively degraded to humic acid, an important organic constituent of soils.

p-Coumarate, which is derived from tyrosine by non-oxidative deamination is a precursor of *p*-hydroxy benzoate which in turn serves as a precursor for the synthesis of ubiquinone, a coenzyme which functions in the electron transport system in mitochondria. Similarly, tyrosine can be converted via a transamination reaction into *p*-hydroxyphenylpyruvate and then into homogentisic acid and thus serves as a precursor of the aromatic nucleus of plastoquinone, a molecule which functions in photosynthetic electron transport in chloroplasts.

* Glucosylation may occur at one or both of these hydroxyl groups to form the water-soluble anthocyanins.

Alkaloids

The term alkaloid is applied to nitrogen-containing molecules belonging to one of the largest and most diverse families of naturally occurring compounds which usually contain nitrogen as part of a heterocyclic system. These compounds are grouped together not because of their ring size or type but because of the presence of a basic nitrogen atom in their structures. It has been estimated that in excess of 6,000 compounds with alkaloid-like properties are present in nature, most of them being found in the Angiospermae and only a few in the Gymnospermae. Most alkaloids have a disturbing effect on the nervous system of animals and many of the 'active principles' of plant extracts which have been used as medicines over many centuries have been shown to be alkaloids. Several amino acids can be considered to be precursors for the biosynthesis of the majority of alkaloids, these amino acids being lysine, ornithine, tryptophan, phenylalanine and tyrosine. However, despite the large number of alkaloids known, information on the synthesis of most of these compounds is very limited.

Many alkaloids are derived directly from aromatic amino acids with tyrosine and phenylalanine serving as precursors of a large group of alkaloids which are characterised by the phenylethylamine or isoquinoline ring structures:

Phenylethylamine   Isoquinoline

The Mannich reaction, in which an amine and an aldehyde (probably via a Schiff-base) react with a nucleophilic carbon such as that of an enolate ion, is involved in the formation of many alkaloids:

Enolate ion   Aldehyde   Amine   (Mannich reaction)

Amino acids can thus provide both of the major substrates for alkaloid biosynthesis since amines and aldehydes can be formed by

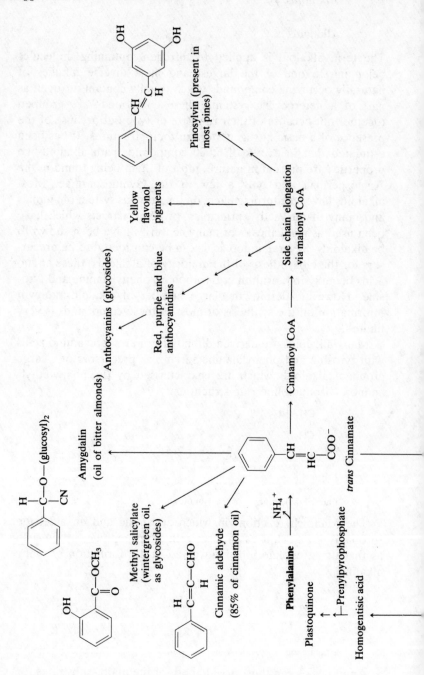

Pinosylvin (present in most pines)

Side chain elongation via malonyl CoA

Yellow flavonol pigments

Anthocyanins (glycosides)

Red, purple and blue anthocyanidins

Cinnamoyl CoA

Amygdalin (oil of bitter almonds)

Methyl salicylate (wintergreen oil, as glycosides)

Cinnamic aldehyde (85% of cinnamon oil)

trans Cinnamate

Phenylalanine

Plastoquinone

Prenylpyrophosphate

Homogentisic acid

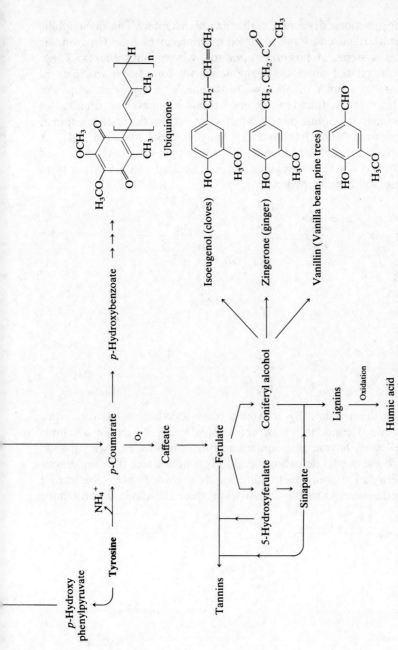

Fig. 3.3 Tyrosine and phenylalanine as precursors of some secondary plant metabolites

the reactions described earlier in this chapter. The nucleophilic centre required for the Mannich reaction can be found in aromatic ring systems at positions *para* to hydroxyl substituents. These hydroxylated aromatic components are found to participate in many reactions of alkaloid biosynthesis, a typical example of which is to be found in the biosynthesis of papaverine, an alkaloid found in the opium poppy, which derives from the condensation of dopamine and 3,4-dihydroxyphenylacetaldehyde (Fig. 3.4). A simpler form of the Mannich reaction is to be seen in the proposed pathway for the biosynthesis of the alkaloid pellotine from tyrosine and acetaldehyde:

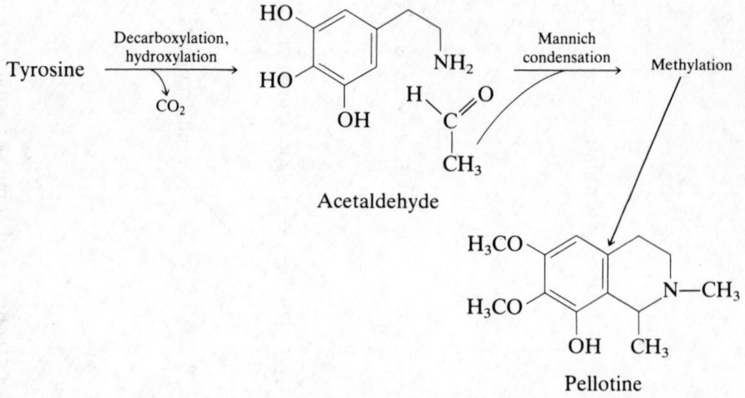

Acetaldehyde

Pellotine

In this reaction hydroxylated phenylethylamine (derived from tyrosine) condenses with acetaldehyde to produce an alkaloid, pellotine, having an isoquinoline ring structure. In barley plants, tyrosine can be decarboxylated to tyramine which then undergoes methylation using methionine or its activated form, S-adenosyl methionine, as methyl donor to produce the alkaloid hordenine:

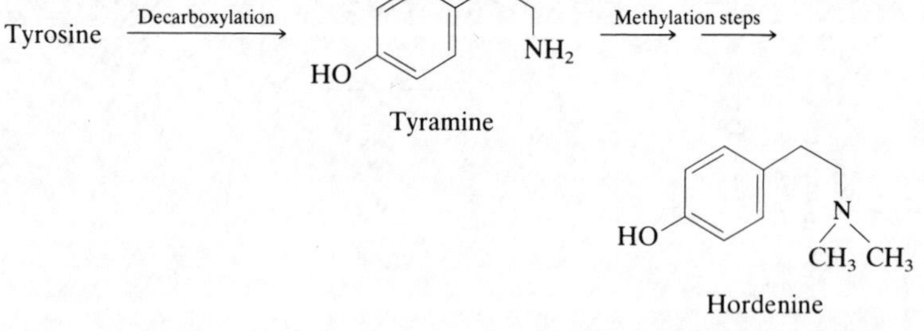

Tyramine

Hordenine

In addition to the Mannich reaction, other more complex alkaloids can be synthesised by both inter- and intramolecular

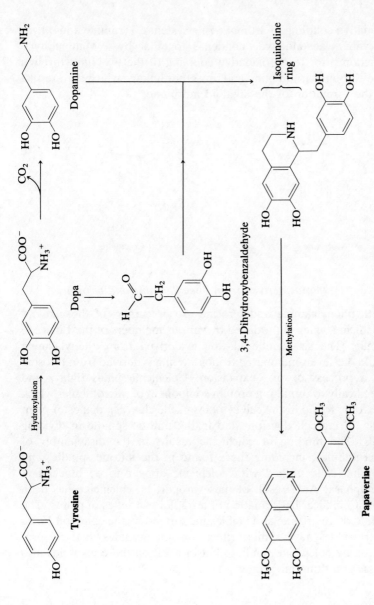

Fig. 3.4 The Mannich reaction in the formation of papaverine

oxidative coupling of aromatic ring systems. Tyramine and dihydroxyphenylacetaldehyde condense together by a Mannich-type reaction prior to an oxidative coupling of the two rings through one carbon–carbon bond and one ether linkage which then results in the formation of the alkaloid morphine.

Morphine   Colchicine

Alkaloids derived from phenylalanine and tyrosine

Both phenylalanine and tyrosine are precursors of the alkaloid colchicine which is produced by various members of the Liliaceae family. The six-membered ring is derived from phenylalanine while the seven-membered tropolone ring is derived from tyrosine by a process of ring expansion. Colchicine binds tightly and specifically to tubulin, a protein component of microtubules which are to be found in the cell cytoplasm, the cleavage planes of plant cells during cell division, and in the mitotic spindle in dividing cells. The binding of colchicine results in the disassembly of microtubules, including those found in the mitotic spindle, and dividing cells treated with colchicine appear to be blocked at metaphase. As a result of these properties, colchicine has been used to induce the formation of tetraploid varieties of plants. The alkaloids vincristine and vinblastine formed by the common *Vinca* (periwinkle) have similar effects on microtubules to those produced by colchicine binding, but in addition these alkaloids also possess antitumour properties.

Alkaloids derived from tryptophan

Tryptophan is a precursor for the biosynthesis of several alkaloids. Hydroxylation of tryptophan produces 5-hydroxy-tryptophan which in turn can be metabolised further to produce a variety of products including the hallucinogenic material psilocybine in the mushroom *Psilocybe aztecorum*:

Tryptophan $\xrightarrow{\text{Hydroxylation}}$ 5-Hydroxytryptophan $\xrightarrow[\text{CO}_2]{\text{Decarboxylation}}$

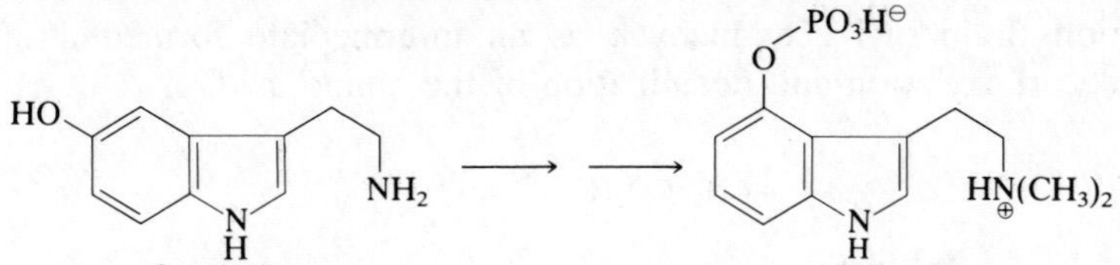

Serotonin
(5-hydroxytryptamine)

Psilocybine

Two other alkaloids which are known to be derived from tryptophan are gramine, an alkaloid found in barley, and lysergic acid (ergot) whose synthesis involves an initial condensation of an isopentenyl group on the indole ring of tryptamine. Conversion of the carboxyl substituent of the aromatic ring system of lysergic acid to a diethylamide ($-CO.N(C_2H_5)_2$) substituent results in the production of a powerful hallucinogenic drug LSD.

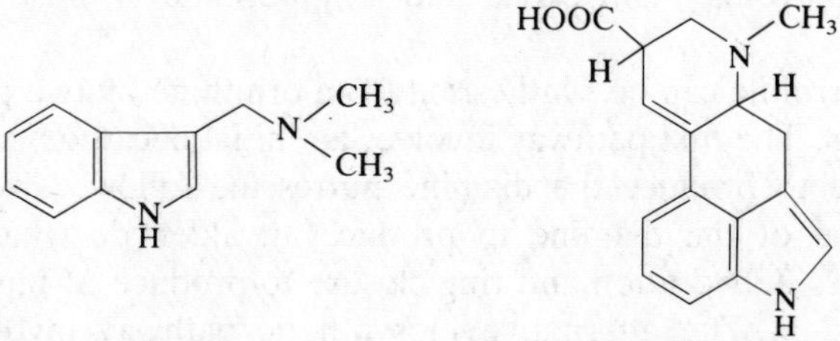

Gramine

Lysergic acid

Tryptophan also contributes a part of the molecule of the alkaloid reserpine found in *Rauwolfia*. The medical interest in this alkaloid results from its ability to cause a lowering of blood pressure and an ability to deplete nervous tissues of serotonin, dopamine and noradrenaline. Reserpine also contains a benzene ring which is derived from tryptophan via a ring expansion process.

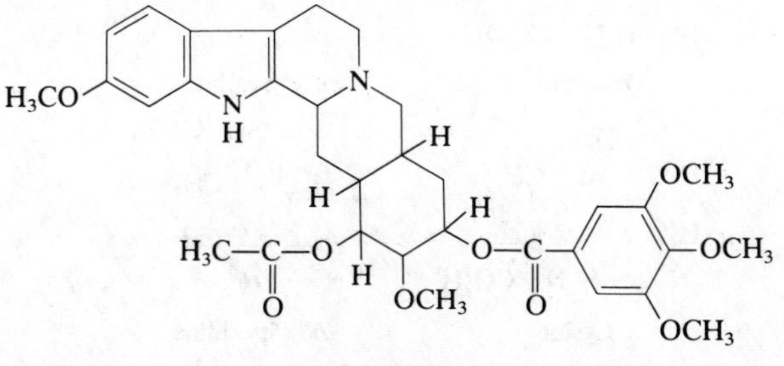

Reserpine

Tryptophan is considered to be the precursor of the plant hormone indole-3-acetate (auxin), although the precise route of synthesis of this hormone has yet to be confirmed. The metabolic route from tryptophan to indole-3-acetate may involve the production of indolyl acetaldehyde as an intermediate formed after decarboxylation and deamination of the amino acid.

$$\text{—CH}_2.\text{CO}_2\text{H}$$

Indole-3-acetic acid

Alkaloids derived from lysine and ornithine

The basic amino acids ornithine and lysine are considered to be the precursors of Δ'-pyrroline and Δ-piperidine respectively, these cyclic compounds then serving as precursors in the synthesis of several alkaloids. The structural relationship between ornithine and Δ'-pyrroline, and lysine and Δ'-piperidine is illustrated in Fig. 3.5.

Δ'-pyrroline can be synthesised from ornithine by two possible pathways. The first pathway involves an initial decarboxylation of ornithine to produce the diamine putrescine followed by transamination of the diamine to produce an aldehyde which then undergoes dehydration and ring closure to produce Δ'-piperidine from lysine. The alternative biosynthetic pathway involves an initial transamination of ornithine to produce glutamic γ-semialdehyde which undergoes ring closure via a dehydration reaction before a final decarboxylation step produces Δ'-pyrroline. Similarly Δ'-piperidine can be synthesised from lysine in the same

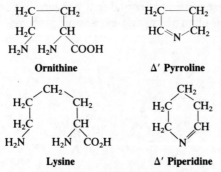

Fig. 3.5 Structural relationship between ornithine and
Δ'-pyrroline, and lysine and Δ'-piperidine

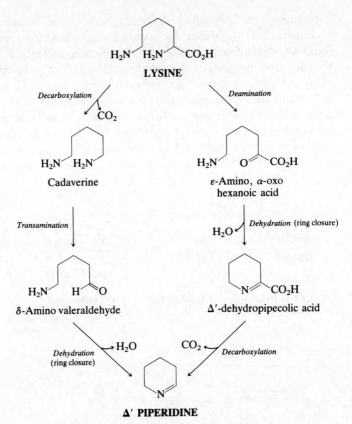

Fig. 3.6 Synthesis of Δ'-piperidine from lysine

way by an analogous series of reactions, the only difference being that an initial deamination of lysine (rather than a transamination as in the case of ornithine) produces an ε-amino, α-keto acid instead of a semialdehyde. These two routes for the synthesis of Δ'-piperidine from lysine are shown in Fig. 3.6. An example of a piperidine alkaloid is anabasine, formed from Δ'-piperidine and nicotinic acid.

Anabasine   Nicotine

Nicotinic acid can also condense with Δ'-pyrroline to produce nicotine, the nicotinic acid being synthesised from glyceraldehyde-3-phosphate and aspartic acid via quinolinic acid formation and subsequent reactions which make up the so-called nicotinic acid cycle (Fig. 3.7). Evidence for such a cycle of reactions has been found in plant tissues. The nicotinic acid mononucleotide formed by the condensation of quinolinic acid and 5'-phosphoribosyl-1-pyrophosphate can in turn either condense with ATP to form NAD and then via transamination, hydrolysis and deamination steps in the nicotinic acid cycle be converted to nicotinic acid, or alternatively lose a ribose-5-phosphate moiety and be converted directly to nicotinic acid. When nicotinic acid is utilised in alkaloid biosynthesis it must first be decarboxylated and enzymes respon-

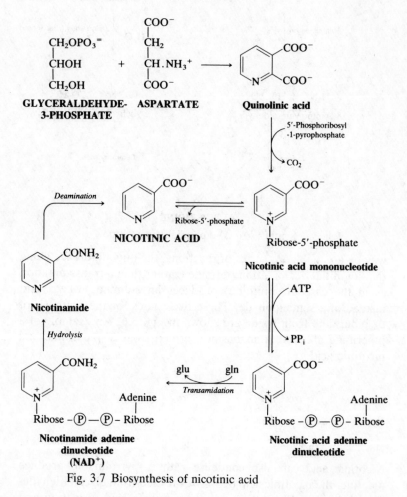

Fig. 3.7 Biosynthesis of nicotinic acid

sible for this reaction have been shown to be present in tobacco roots.

Ethylene biosynthesis

Ethylene (C_2H_4), the simplest plant hormone, is capable of initiating fruit development and regulating numerous other processes in plants. The sulphur-containing amino acid methionine is the biological precursor of ethylene in all higher plant tissues and the biosyntheic pathway for ethylene production is outlined in Fig. 3.8.

The fates of the various carbon atoms of methionine utilised for ethylene biosynthesis in the cell are shown in the diagram:

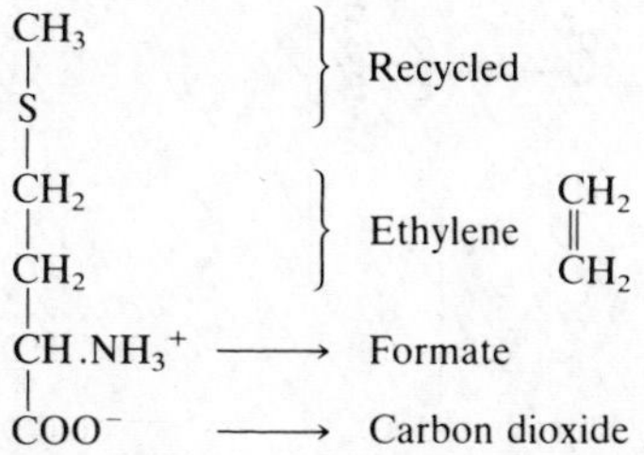

$$
\begin{array}{l}
CH_3 \\
| \\
S
\end{array} \Bigg\} \text{ Recycled}
$$

$$
\begin{array}{l}
CH_2 \\
| \\
CH_2
\end{array} \Bigg\} \text{ Ethylene} \quad
\begin{array}{l}
CH_2 \\
\| \\
CH_2
\end{array}
$$

$$
\begin{array}{l}
| \\
CH.NH_3^+ \longrightarrow \text{ Formate} \\
| \\
COO^- \longrightarrow \text{ Carbon dioxide}
\end{array}
$$

As can be seen, it is the C-3 and C-4 atoms of methionine which are conserved in ethylene biosynthesis. The sulphur atom of the methionine molecule is also retained and recycled via methylribose to re-form methionine. If the sulphur atom was not recycled then there would be a danger that methionine concentrations, which are normally low in plant tissues, would become limiting. In this biosynthetic pathway (Fig. 3.8) methionine is initially converted into S-adenosylmethionine, an 'activated' form of methionine containing an excellent leaving group (the positively charged sulphonium group). S-adenosyl methionine undergoes a typical γ-elimination (1,3-elimination) reaction catalysed by ACC synthase, an enzyme which is thought to contain pyridoxal phosphate as coenzyme, producing methyl thioadenosine which enters the sulphur recycling loop and 1-aminocyclopropane-1-carboxylic acid (ACC) which can then be oxidised to ethylene. This final oxidation step is poorly characterised but the enzyme system responsible appears to be membrane bound and the oxidation process could take place either via a hydroxylation step followed by dehydration or via dehyrogenation of the amino group to produce a nitrenium intermediate which can undergo spontaneous dissociation into ethylene and other products.

76

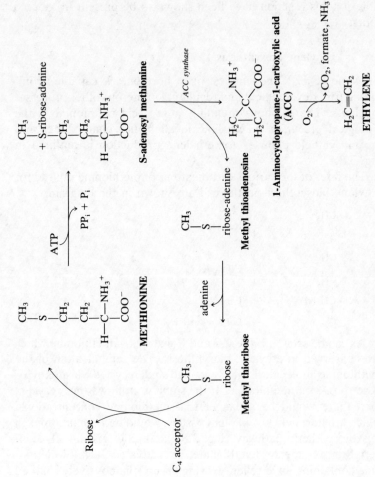

Fig. 3.8 Ethylene biosynthesis

Many of the effects of indole acetic acid (auxin) on plant growth can be attributed to the induction of ethylene production in plant tissues by auxin. Auxin does not control ethylene production via an effect on the reaction converting ACC to ethylene, the enzyme system responsible for catalysing this step being constitutive in most plant tissues except unripe fruit tissue. Neither is the conversion of methionine to S-adenosyl methionine directly affected by auxin. However, auxin does increase the activity of ACC synthase, the enzyme catalysing the rate-limiting step in ethylene production (see Fig. 3.8), most probably via the induction of *de novo* enzyme synthesis. Induction of ACC synthase in this way appears to be the mechanism by which many of the ethylene-inducing stimuli exert their effects in plant tissues.

S-adenosyl methionine

The major methyl donor in most biosynthetic reactions is S-adenosyl methionine (SAM) which is formed by the transfer of an adenosyl group from ATP to the sulphur atom of the amino acid methionine:

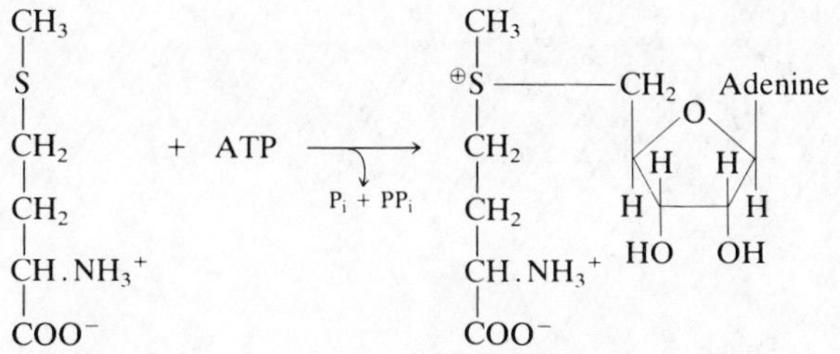

The positive charge on the sulphur atom activates the methyl group of SAM and makes it much more reactive than the corresponding methyl group of 5-methyltetrahydrofolate, another carrier of one-carbon units in the cell. An unusual feature of SAM biosynthesis is that the triphosphate group of ATP is hydrolysed initially into inorganic phosphate and pyrophosphate, the pyrophosphate so produced then being hydrolysed to P_i.

Porphyrin biosynthesis

Porphyrins are comprised of pyrroles which are linked together in a cyclic structure via unsaturated methylene bridges between α-carbon atoms of adjacent pyrrole rings.

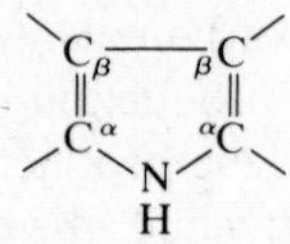

Pyrrole ring

The β-carbon atoms of the pyrrole rings can be linked to a variety of side chains, hence many different porphyrin structures can be produced. Metal ions may also be introduced into the centre of the porphyrin ring structure via bonding to the nitrogen atoms of the pyrrole rings. An example of one class of porphyrins which are of great importance in green plant tissues are the chlorophylls which contain magnesium as the metal ion in the porphyrin ring structure:

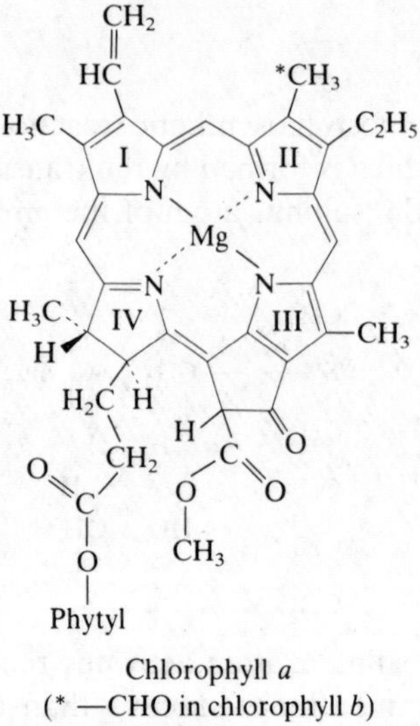

Chlorophyll *a*
(* —CHO in chlorophyll *b*)

Porphyrin biosynthesis can be considered to take place in three stages:

    1 an initial synthesis of δ-aminolaevulinic acid (ALA);
    2 formation of the pyrrole porphobilinogen;
    3 tetrapyrrole formation and cyclisation.

δ-Aminolaevulinic acid synthesis

The first known intermediate which is committed to the pathway for tetrapyrrole biosynthesis is δ-aminolaevulinic acid (ALA) which is synthesised from the amino acid glutamic acid or its oxo acid derivative, 2-oxoglutarate. The possible rates of synthesis of ALA from glutamic acid are outlined in Fig. 3.9 where it can be seen that in two of the routes (B and C) the intact carbon skeleton of 2-oxoglutarate enters ALA while in a third alternative biosynthetic route (A), succinyl CoA and glycine are the substrates for a condensation reaction catalysed by the enzyme δ-aminolaevulinic acid synthetase (ALA synthetase).

The actual pathway by which ALA is synthesised in plants has not been completely characterised. ALA synthetase which catalyses ALA formation in bacteria and animals appears to be absent in greening tissues where large amounts of ALA are required for chlorophyll biosynthesis although this enzyme is thought to be important in the synthesis of cyclic tetrapyrroles in non-green cells and has been shown to be present in soybean callus and cold-stored potatoes. Evidence for ALA synthesis by both pathways (B) and (C) (Fig. 3.9) has been found in leaf tissue but present evidence indicates that pathway (C), involving the production of glutamic-1-semialdehyde from glutamate prior to ALA formation, is the pathway which provides the bulk of the ALA for chlorophyll synthesis in greening plant tissues. This pathway has been demonstrated to operate in greening leaves of barley, maize and pea, in greening cucumber cotyledons and in immature spinach leaves. The reaction sequence appears to be localised in the chloroplasts of plant cells although the enzymes responsible for ALA synthesis are themselves synthesised in the cell cytoplasm, being coded on nuclear genes, and must therefore be translocated specifically into chloroplasts prior to their participation in ALA synthesis.

The different ALA-forming pathways may exist within different subcellular compartments and are probably under separate regulatory control. In most cases which have been studied, the synthesis of ALA has been found to be the rate-limiting step in chlorophyll biosynthesis and the control of ALA synthesis is thus of central importance to the development of photosynthetic competence in plants.

Formation of the pyrrole ring

Porphobilinogen synthesis is initiated by an aldol condensation reaction between two molecules of ALA, followed by a dehydration reaction which introduces a carbon – carbon double bond into the molecule and a transimination reaction which results in ring closure and production of a recognisable pyrrole ring (Fig. 3.10).

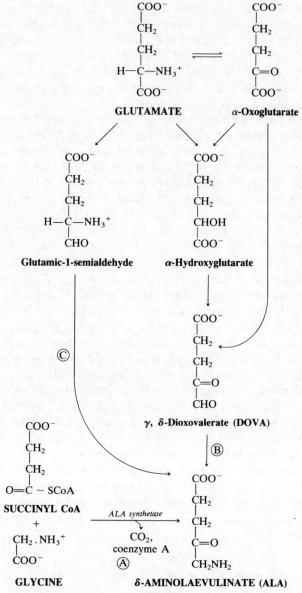

Fig. 3.9 Possible pathways of δ-aminolaevulinate formation in plant tissues

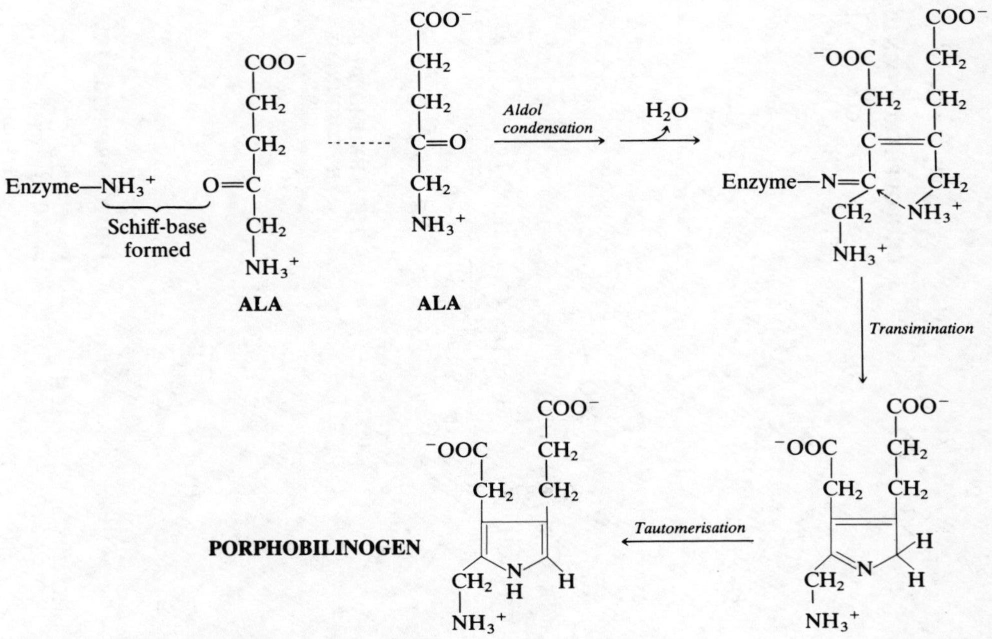

Fig. 3.10 Porphobilinogen synthesis

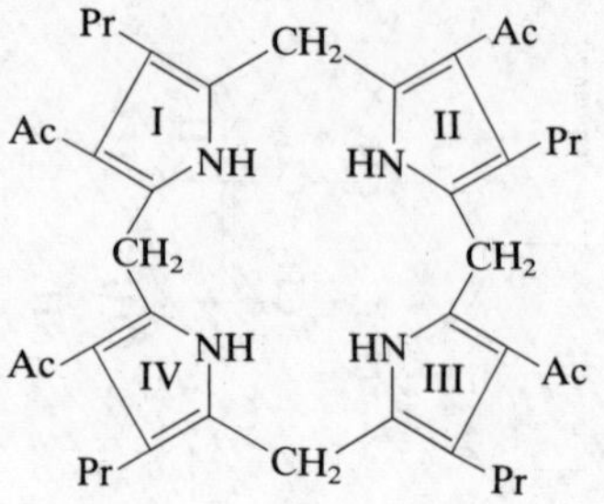

UROPORPHYRINOGEN III

$$\left[\begin{array}{l} Ac = -CH_2.COO^- \\ Pr = -CH_2.CH_2.COO^- \end{array}\right]$$

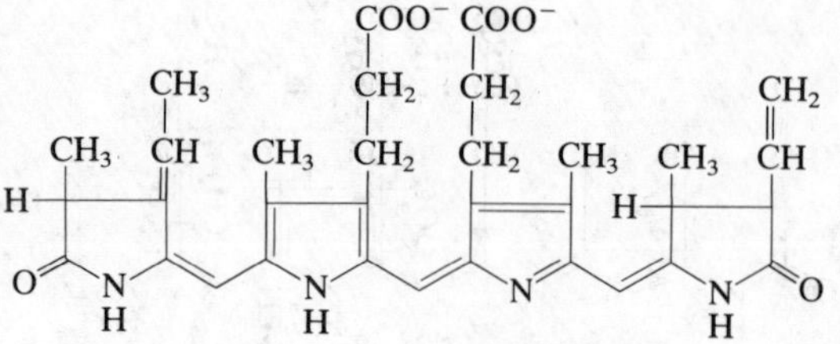

THE CHROMOPHORE OF PHYTOCHROME

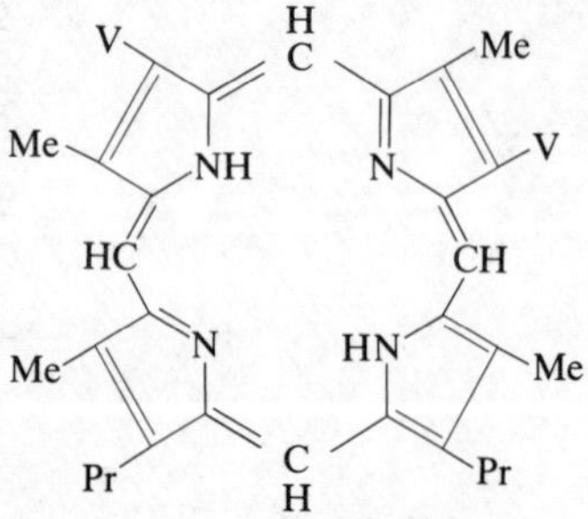

PROTOPORPHYRIN IX

$$\left[\begin{array}{l} Me = -CH_3 \\ Pr = -CH_2.CH_2.COO^- \\ V = -CH = CH_2 \end{array}\right]$$

Fig. 3.11 Tetrapyrroles

A final tautomerisation step results in porphobilinogen production. The initial reaction in porphobilinogen synthesis may involve Schiff-base formation with a carboxyl group on one of the ALA molecules. The enzyme catalysing the synthesis of porphobilinogen from ALA, porphobilinogen synthase, has been detected in plant tissues.

Tetrapyrrole formation and cyclisation

Polymerisation of four porphobilinogen molecules produces a tetrapyrrole and subsequent ring closure results in the formation of either uroporphyrinogen I or uroporphyrinogen III. The latter compound, uroporphyrinogen III, has an altered orientation of pyrrole ring IV from that expected from a linear condensation and cyclisation of four porphobilinogen residues (Fig. 3.11). The porphobilinogen to uroporphyrinogen III conversion is catalysed by uroporphyrinogen I synthase in the presence of a second protein (co-synthase). In the absence of co-synthase, uroporphyrinogen I is produced. Ammonia is eliminated during this reaction:

$$4 \text{ Porphobilinogen} \rightarrow \text{Uroporphyrinogen} + 4NH_3$$

Straight chain as well as cyclic tetrapyrroles are found in plant tissues, one acting as the chromophore of phytochrome (Fig. 3.11). Uroporphyrinogen III then undergoes a series of oxidation and decarboxylation reactions resulting in the production of protoporphyrin IX (Fig. 3.11).

The remaining steps of chlorophyll synthesis from protoporphyrin IX have yet to be elucidated although the first step is possibly the enzyme-catalysed insertion of Mg^{2+} into the molecule followed by a methylation of the proprionyl side chain on ring III:

$$\text{Protoporphyrin IX} \xrightarrow{Mg^{2+}} \text{Mg protoporphyrin IX} \xrightarrow{SAM} \text{Mg protoporphyrin IX methyl ester}$$

Insertion of a ferrous ion in place of Mg^{2+} into protoporphyrin in chloroplasts requires an enzyme, ferrochelatase, which is firmly bound to the chloroplast inner membrane and this chelation of iron gives rise to cytochromes, catalase and peroxidases.

The remaining steps in chlorophyll biosynthesis include saturation of the vinyl group on ring IV, closure of ring V to produce protochlorophyllide *a* which when coupled with phytol, probably utilising phytyl pyrophosphate as phytyl donor, yields chlorophyll *a* (see p. 78), the major pigment of chloroplasts and the centrally important chromophore for photosynthesis in green plants. Chlorophyll *b*, which is nearly always present in green leaves is most probably derived from chlorophyll *a* and its formation involves replacement of the methyl group on ring II by an aldehyde group (p. 78).

Cyanogenic glycosides

The process of HCN production by living organisms is known as cyanogenesis. In plants which possess this capability the cyanogenic substances are of two types, cyanogenic glycosides and cyanogenic lipids, both of which liberate a carbonyl compound and HCN when the sugar or fatty acid moiety is removed. While about 2,000 species of higher plants in over 100 different families are cyanogenic, the number of cyanogenic glycosides and lipids is less than 30. Indeed, there are only four known cyanogenic lipids and these occur in a single family, the Sapindaceae. Examples of the cyanogenic glycosides present in plants are amygdalin (Fig. 3.3) found primarily in seeds of Rosaceae, and dhurrin which can comprise 3–5 per cent of the dry weight of sorghum seedlings.

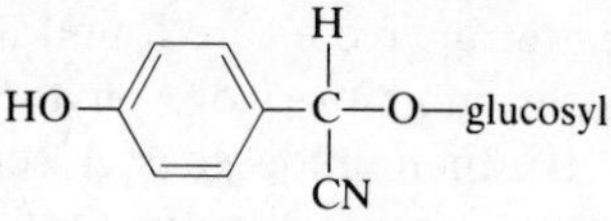

Dhurrin

Cyanogen degradation

The production of HCN from these cyanogenic compounds requires both the cyanogen and the enzymes responsible for its catabolism to be present at the same site in the same tissue. Two enzymes are responsible for HCN production:

1 an initial glycosidase which hydrolyses the cyanogenic glycoside into sugar and α-hydroxynitrile (cyanohydrin) components;

2 a hydroxynitrile lyase which liberates HCN and an aldehyde or ketone.

This process can be represented by the general reaction sequence:

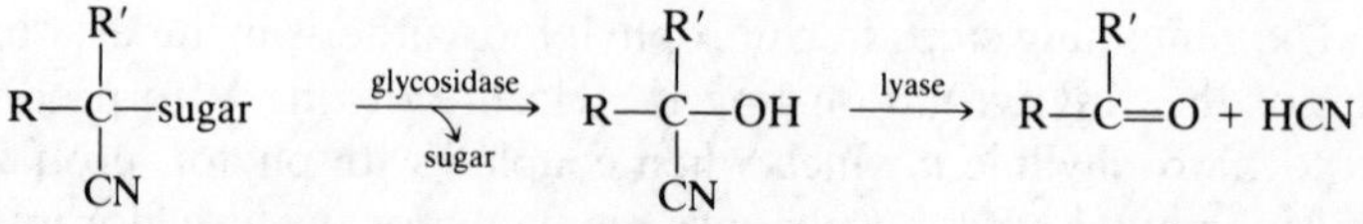

Cyanogenic glycoside   Cyanohydrin   Aldehyde or ketone
                                aglycone

In plant tissues, the cyanogenic glycosides may normally be physically separated from their hydrolytic enzymes unless the tissue is damaged or crushed. In young green sorghum leaf tissue virtually all the dhurrin is located in epidermal protoplasts while the glycosidase and lyase enzymes responsible for its degradation are found only in mesophyll tissue. Often the cyanogenic process is set in action when plant tissue is chewed during ingestion by animals resulting in poisoning of the animal by plant products which contain cyanogenic substrates.

Cyanogen biosynthesis

The majority of known cyanogenic glycosides are synthesised from the amino acids tyrosine, phenylalanine, valine, leucine, or iso-leucine via a reaction sequence which appears to be membrane localised. The reactions of the the pathway can be represented by the sequence shown in Fig. 3.12.

The amino acid is first converted to an oxime in a reaction which involves α-decarboxylation, the oxime then being dehydrated to produce the nitrile. The nitrile acquires a hydroxyl group

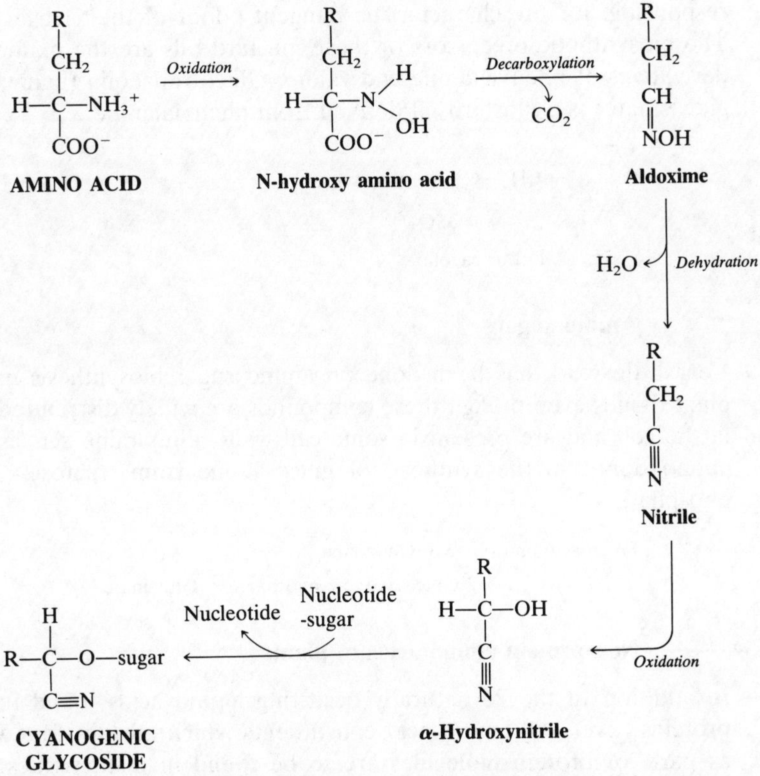

Fig. 3.12 Biosynthesis of cyanogenic glycosides

and is subsequently glycosylated in a reaction which probably involves a nucleotide sugar derivative. The few cyanogenic glycosides which do not use the previously mentioned amino acids as precursors appear to be formed from the non-protein amino acid cyclopentenyl glycine and an example of these cyclopentenoid glycosides is gynocardin:

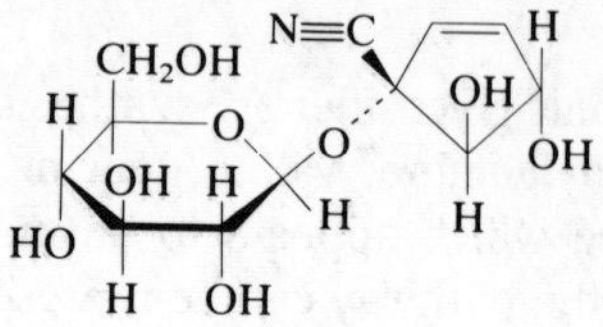

Gynocardin

Mustard oils

Mustard oil glucosides occur widely in the Cruciferae and are responsible for the characteristic pungent odour of their leaves. The biosynthetic precursors of these mustard oils are the oxime derivatives of phenylalanine and valine. Glucotropaeolin (benzyl glucosinate) is a mustard oil derived from phenylalanine:

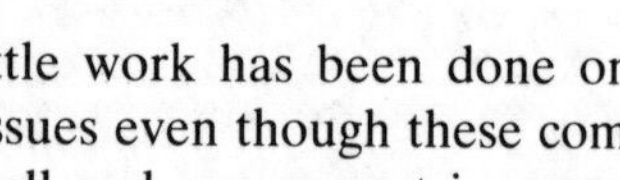

Glucotropaeolin

Amino sugars

Very little work has been done on amino sugar biosynthesis in plant tissues even though these compounds are widely distributed in the cell and are present in some cell walls. Glutamine acts as amino donor in the synthesis of glucosamine from fructose-6-phosphate:

Fructose-6-phosphate + Glutamine

$\rightarrow$ 2-Glucosamine-6-phosphate + Glutamate

Non-protein amino acids of plants

In addition to the 20 naturally occurring amino acids found in proteins, over 200 amino acid constituents which are not found as parts of protein molecules are to be found in plant tissues. Discussion of the metabolism of these compounds is outside the scope of this chapter but the reader is referred to the excellent reviews by Lea (1978) and Fowden, Lea and Bell (1979) listed in the reading list below.

Suggestions for further reading

Adams, D. O. & Yang, S. F. (1981) Ethylene the gaseous plant hormone: mechanism and regulation of biosynthesis, *Trends in Biochemical Sciences*, **6**, 161–3.

Beale, S. I. (1978) δ-Aminolevulinic acid in plants: its biosynthesis regulation and role in plastid development, *Ann. Rev. Plant Physiol.*, **29**, 95–120.

Conn, E. E. (1980) Cyanogenic compounds, *Ann. Rev. Plant Physiol.*, **31**, 433–51.

Fowden, L., Lea, P. J. & Bell, E. A. (1979) The non-protein amino acids of plants, *Advances in Enzymology*, **50**, 117–75.

Lea, P. J. (1978) Biosynthesis of unusual amino acids, in H. R. V. Arnstein (ed.) *International Review of Biochemistry: Amino Acid and Protein Synthesis II*, Vol. 18. University Park Press: Baltimore.

Waller, G. R. & Nowacki, E. K. (1978) *Alkaloid Biosynthesis and Metabolism in Plants*. Plenum.

4

Purines, pyrimidines and their derivatives

Once nitrogen has been incorporated into one of the amino acids glycine, glutamine, aspartate or aspargine then it is possible for it to be incorporated into either purine or pyrimidine bases. These nitrogenous bases are components of the nucleic acid molecules in the cell, i.e. DNA (deoxyribonucleic acid, the carrier of genetic information in the cell) and RNA (ribonucleic acid, the molecules involved in the expression of genetic information in the cell). Also, purine and pyrimidine derivatives are involved in several metabolic processes such as carbohydrate and lipid metabolism and as component parts of pyridine and flavin nucleotide coenzymes which in turn are involved in many of the oxidation – reduction reactions of cellular metabolism.

Pyrimidine bases

The pyrimidine bases are derivatives of the parent compound pyrimidine whose basic structure has a six-membered ring with nitrogen atoms at positions 1 and 3.

Pyrimidine   Cytosine   Uracil   Thymine (5-methyl uracil)

The principal pyrimidine derivatives found in nucleic acids are cytosine (found in DNA and RNA), uracil (RNA) and thymine (DNA) but other hydroxyl, amino and methyl-substituted derivatives can occur in small amounts in nucleic acids, e.g. 4-thiouracil and 5-methyl cytosine are found in transfer RNA.

Purine bases

The purine bases are derivatives of the parent compound purine in which a five-membered imidazole ring has fused with a pyrimidine to form a nine-membered purine ring.

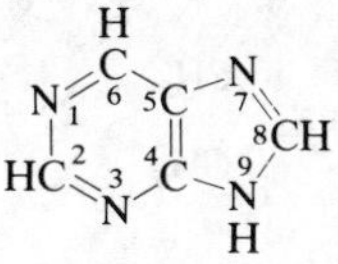

Purine

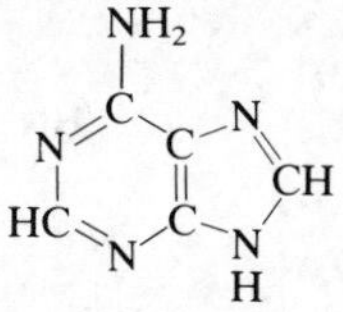

Adenine

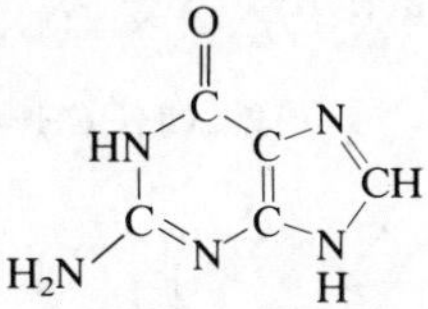

Guanine

The same purines, adenine and guanine, are found in both DNA and RNA but other purines which are not found in nucleic acids are to be found in some plants as the free bases. These comprise mainly xanthine and methylated xanthine derivatives such as theophylline and caffeine found in tea and coffee leaves and theobromine found in cocoa. The physiological role of these xanthine derivatives is unclear, but theophylline and caffeine have been used in biochemical studies on the action of animal hormones since these two derivatives are potent inhibitors of the phosphodiesterase enzyme which degrades cyclic AMP, a second messenger in the action of many animal hormones.

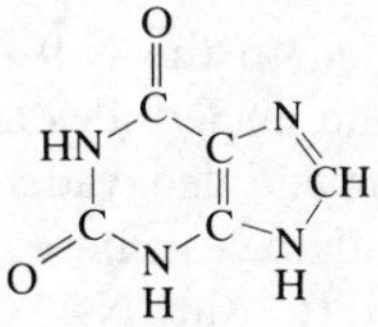

Xanthine

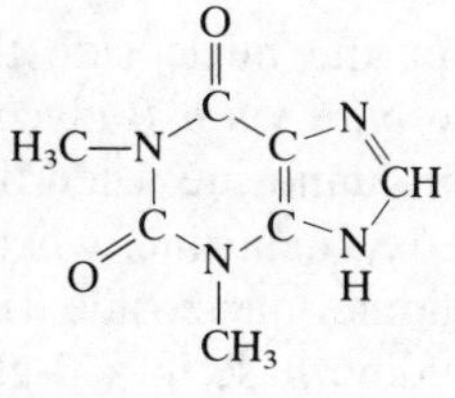

Theophylline (1,3-dimethyl xanthine) found in leaves of *Thea sinensis*

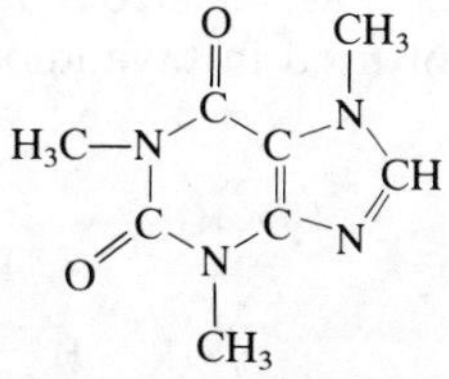

Caffeine (1,3,7-trimethyl xanthine) found in leaves of *Thea sinensis* and *Coffea arabica*

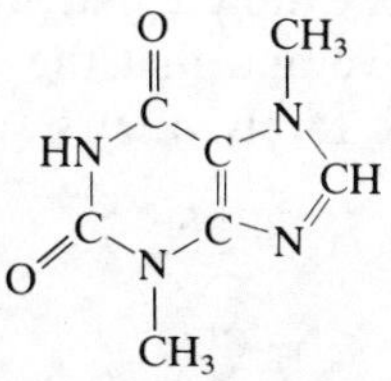

Theobromine (3,7-dimethyl xanthine) found in *Theobroma cacao*

Methylated derivatives of adenine and guanine can be found as 'minor' bases in nucleic acids and adenine derivatives containing the isopentenyl group have been found in transfer RNA molecules. An interesting finding concerning these isopentenyl-substituted adenine compounds is that they possess cytokinin activity.

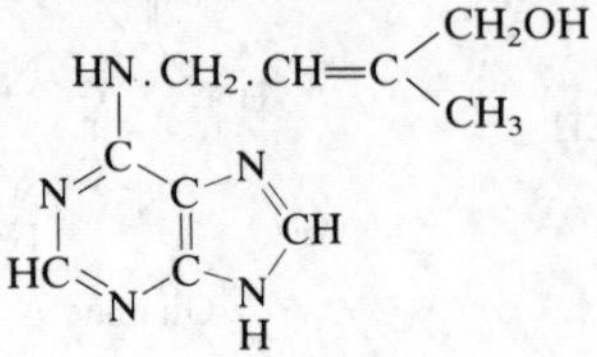

Zeatin
6-(4-hydroxy-3-methylbut-2-enyl) amino purine
The first natural cytokinin isolated from corn (*Zea mays*) kernels

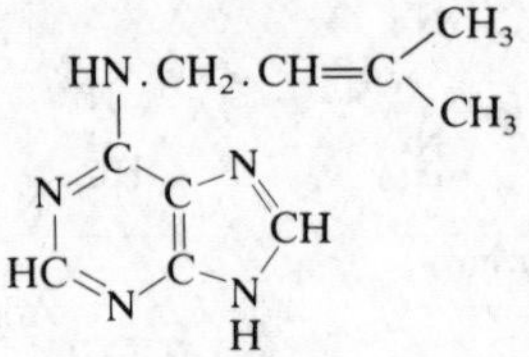

Isopentenyl adenine
6-($\gamma\gamma$-dimethylallyl) amino purine

Nucleosides and nucleotides

The purine and pyrimidine bases seldom occur in the free state in cells but can be found complexed with a sugar, usually ribose or deoxyribose to form a nucleoside, or with a pentose-phosphate sugar to form a nucleotide.

Nucleosides

In a purine nucleoside the sugar is attached to the N-9 of the purine base via a β-glycosidic link giving the nucleosides adenosine or guanosine when the sugar is D-ribose and deoxyadenosine or deoxyguanosine when the sugar is 2-deoxy-D-ribose. In a pyrimidine nucleoside the sugar is linked to the N-1 of the pyrimidine base in a β-glycosidic link to give uridine or cytidine when uracil or cytosine is attached to ribose or deoxycytidine or thymidine when cytosine or thymine is attached to deoxyribose. It should be noted that uracil does not occur as the deoxyribose derivative and that the deoxy prefix is not used in thymidine as thymine rarely occurs attached to ribose.

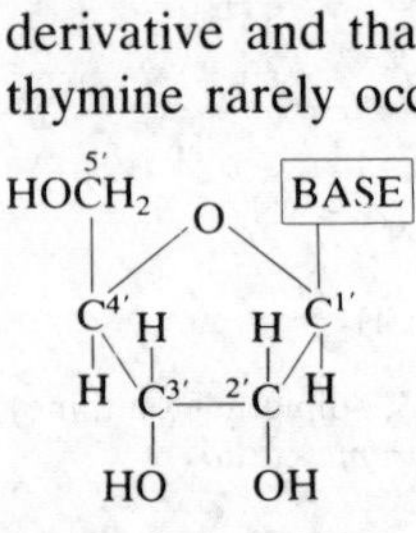

Ribonucleoside

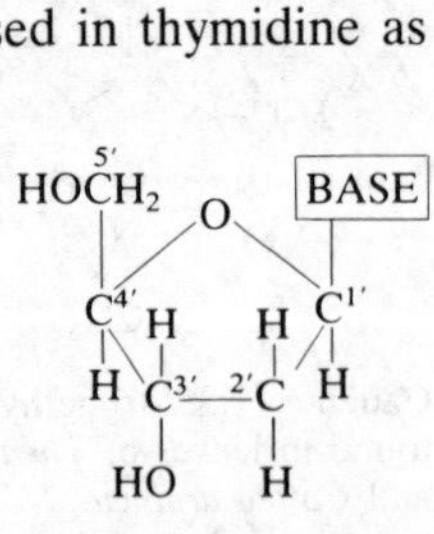

Deoxyribonucleoside

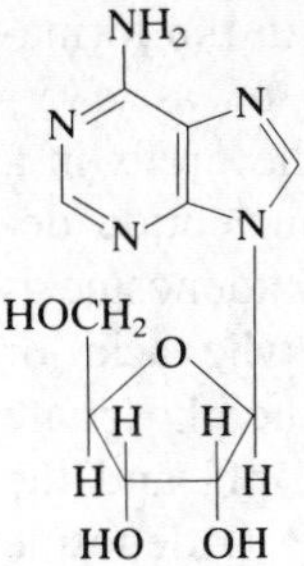

Adenosine
(ribonucleoside)

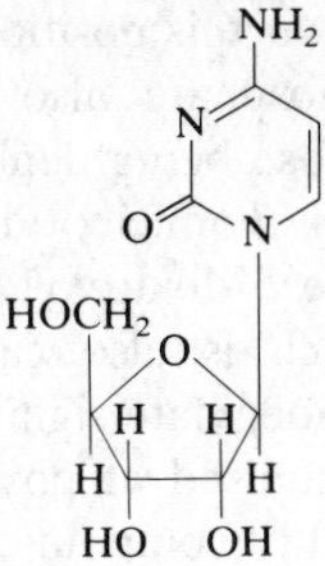

Cytidine
(ribonucleoside)

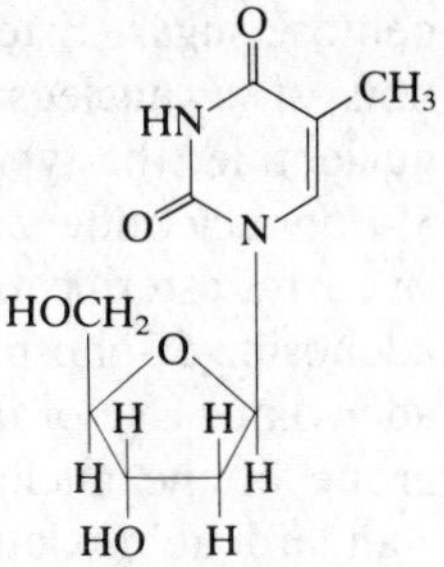

Thymidine
(deoxyribonucleoside)

Whereas free bases are seldom found in significant amounts in plant cells, nucleosides can be detected in the free state along with other base – sugar complexes the nature of which depends upon the plant species. *Croton tiglium* contains the free nucleoside crotonoside consisting of the purine isoguanine and the pentose sugar ribose, while seed of the *Vicia* species contains a glucoside vicin in which the pyrimidine divicin is linked via an oxygen substituent to the sugar. In addition zeatin has been reported to occur in plant cells as the riboside.

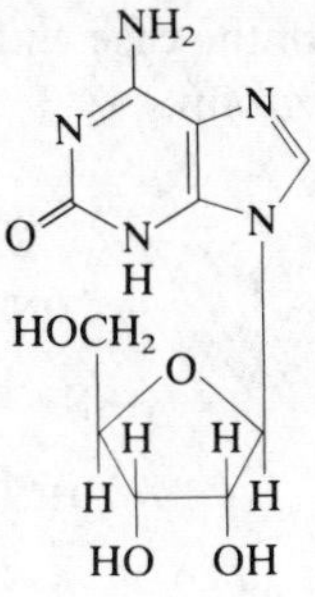

Crotonoside

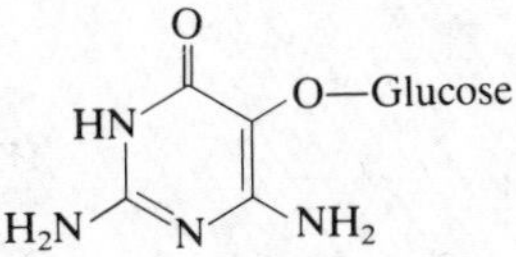

Vicin

Nucleotides

The addition of a phosphate group to the sugar residue of a nucleoside results in the production of a nucleotide, the esterification of the phosphate to the sugar being possible at the 2'-, 3'- and 5'-hydroxyl groups of the ribose sugar moiety of the nucleosides and the 3'- and 5'-hydroxyl groups of a deoxyribose moiety. A primed (') number represents an atom of the pentose or deoxypentose sugar whereas an unprimed number represents an atom of the purine or pyrimidine ring. The most common site of esterification is the hydroxyl group attached to the C-5' of the

pentose sugar. Esterification at this position results in the formation of a nucleoside 5'-phosphate also referred to as a 5'-
nucleotide, the type of pentose being denoted by the prefix in a
5'-*ribo*nucleotide or 5'-*deoxyribo*nucleotide. The nucleotide derived by esterification of the 5'-hydroxyl group of adenosine is
adenosine-5'-phosphate which is also called adenylic acid or
adenylate. Use of the term adenylate signifies that the phosphate
group of the nucleotide is ionised at physiological pH and the
standard abbreviation for this compound is AMP (*A*denosine
*Mono*Phosphate). Other major 5'-nucleotides found in the cell are
guanylate (GMP), cytidylate (CMP) and uridylate (UMP) while
the major 5'-deoxyribonucleotides are deoxyadenylate (dAMP),
deoxyguanylate (dGMP), deoxycytidylate (dCMP) and deoxy-
thymidylate (dTMP), although this latter deoxyribonucleotide is
commonly referred to as thymidylate (TMP) while the rather rare
ribonucleotide containing ribose, phosphate and thymine is
termed ribothymidylate. Although in nature the prominent free
nucleotides are the 5'-phosphate esters, degradation of nucleic
acids in the cell results in the production of 3'-phosphate derivatives. Since ribonucleosides have three free hydroxyl groups, then
in addition to the 3'- and 5'-phosphate esters an additional
2'-phosphate ester may be formed. This is not the case with deoxy-
ribonucleosides which lack the 2'-hydroxyl group.

Adenosine-5'-phosphate   Thymidine-5'-phosphate   Guanosine-3'-phosphate
(AMP)   (TMP)   (3'-GMP)

The ribonucleoside- and deoxyribonucleoside-5'-monophos-
phates can be further phosphorylated at the 5'-position to yield
5'-diphosphates, e.g. adenosine-5'-diphosphate (ADP) and 5'-
triphosphates, e.g. adenosine-5'-triphosphate (ATP). Also, adeno-

Table 4.1 Nomenclature of bases, nucleosides and nucleotides

			5′-Monophosphates	
			---	---
Base	Ribonucleoside	Deoxyribonucleoside	Ribonucleotide	Deoxyribonucleotide
Adenine	Adenosine	Deoxyadenosine	Adenylate (AMP)	Deoxyadenylate (dAMP)
Guanine	Guanosine	Deoxyguanosine	Guanylate (GMP)	Deoxyguanylate (dGMP)
Cytosine	Cytidine	Deoxycytidine	Cytidylate (CMP)	Deoxycytidylate (dCMP)
Uracil	Uridine		Uridylate (UMP)	
Thymine		Thymidine		Thymidylate (TMP)

sine-5′-and guanosine-5′-tetraphosphates have been found in cells. The nomenclature of bases, nucleosides and nucleotides is summarised in Table 4.1. It is also possible to form cyclic nucleotides, e.g. 2′,3′-AMP, 3′,5′-AMP, 3′,5′-GMP, and these have been detected in animal, plant and bacterial cells.

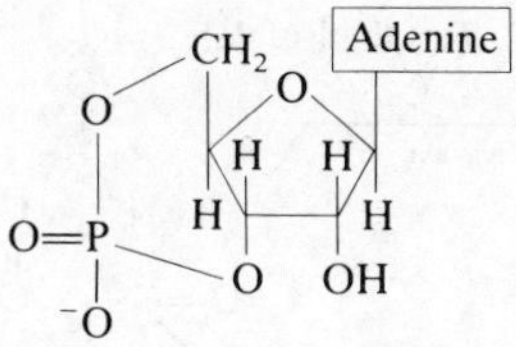

Adenosine-3′,5′-cyclic monophosphate
(cyclic AMP, cAMP)

Biosynthesis of purines

Purines are synthesised in the cell in the form of their nucleoside monophosphates and the origin of the carbon and nitrogen atoms of the purine ring is as shown below:

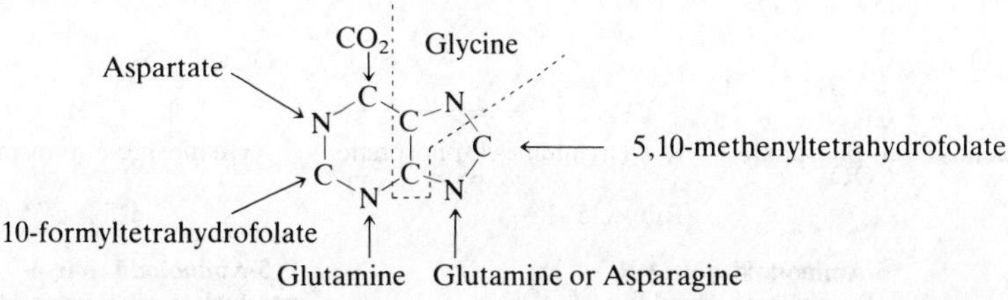

The series of reactions by which purines are thought to be synthesised in higher plants is described in Fig. 4.1, the evidence

94

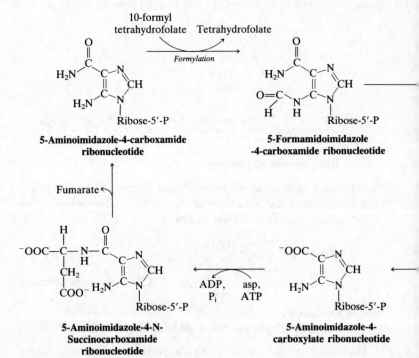

Fig. 4.1 Biosynthesis of the purine ring

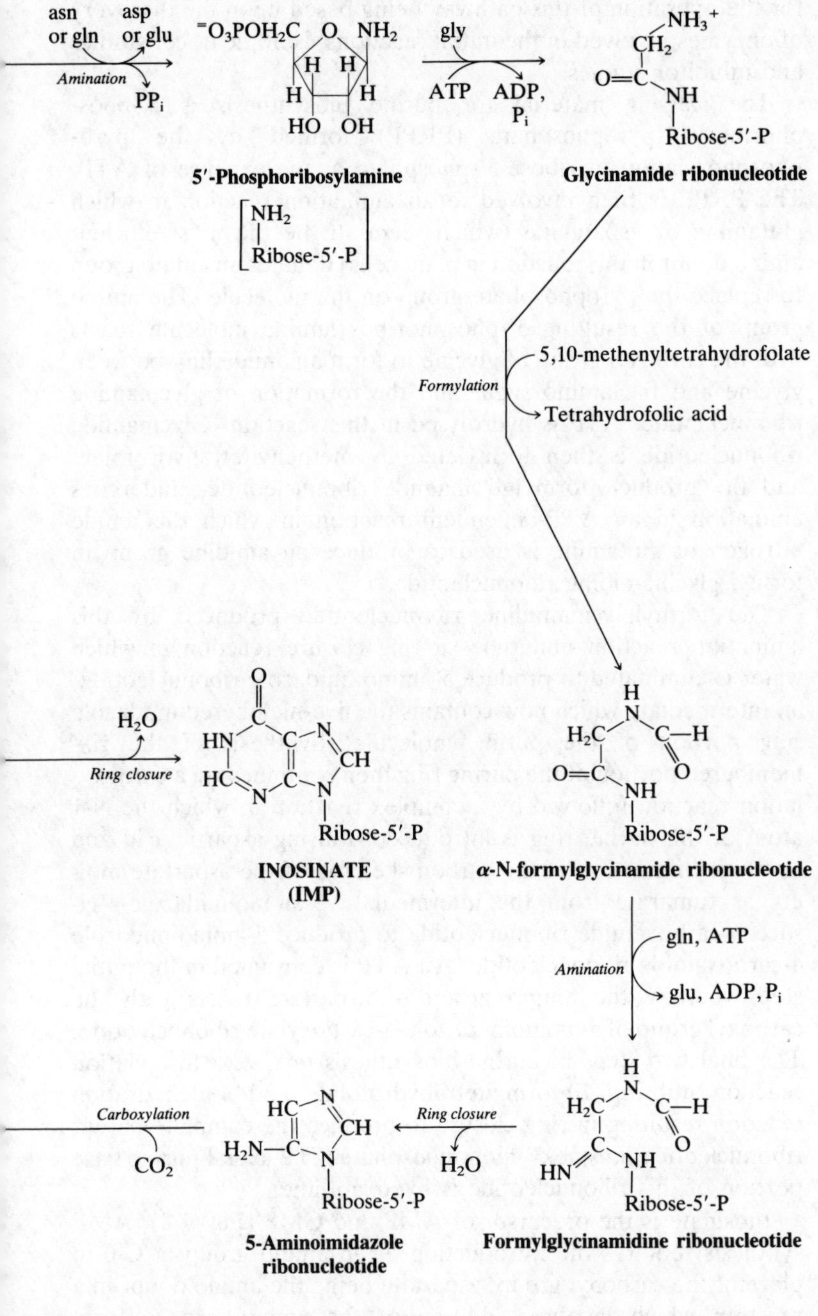

asn or gln / asp or glu
Amination → PP$_i$

$^-O_3POH_2C$ O NH$_2$
H H
H H
HO OH

5′-Phosphoribosylamine

$$\begin{bmatrix} NH_2 \\ | \\ Ribose\text{-}5'\text{-}P \end{bmatrix}$$

gly
ATP → ADP, P$_i$

NH$_3{}^+$
CH$_2$
O=C
NH
Ribose-5′-P

Glycinamide ribonucleotide

Formylation — 5,10-methenyltetrahydrofolate → Tetrahydrofolic acid

H$_2$O
Ring closure →

O
HN—C—C—N
HC—C—N—CH
N
Ribose-5′-P

INOSINATE (IMP)

H
H$_2$C—C—H
O=C
NH
O
Ribose-5′-P

α-N-formylglycinamide ribonucleotide

gln, ATP
Amination → glu, ADP, P$_i$

H
H$_2$C—N—C—H
C
HN NH
O
Ribose-5′-P

Formylglycinamidine ribonucleotide

Carboxylation
CO$_2$

HC—N
H$_2$N—C—N—CH
Ribose-5′-P

5-Aminoimidazole ribonucleotide

Ring closure
H$_2$O

for the operation of this pathway being based upon the discovery of enzymes involved in the initial reactions, isotopic tracer studies and inhibitor studies.

The starting material for purine biosynthesis is 5'-phosphoribosyl-1-pyrophosphate (PRPP) formed by the pyrophosphorylation of ribose-5'-phosphate at the expense of ATP. The PRPP is then involved in an amination reaction in which glutamine or asparagine (which seems to be the most efficient amino donor in this reaction in plant cells) donates an amino group to replace the pyrophosphate group on the molecule. The amino group of the resulting 5'-phosphoribosylamine molecule reacts with the carboxyl group of glycine to form an amide link between glycine and the amino sugar and the formation of glycinamide ribonucleotide. ATP is hydrolysed in this reaction. Glycinamide ribonucleotide is then formylated by methenyltetrahydrofolate and the product, formylglycinamide ribonucleotide, undergoes amination in an ATP-dependent reaction in which the amide nitrogen of glutamine is used to produce an amidine group in formyl glycinamidine ribonucleotide.

The formylglycinamidine ribonucleotide produced by this amination reaction undergoes a ring closure reaction in which water is eliminated to produce 5-amino imidazole ribonucleotide, an intermediate which now contains the five-membered imidazole ring portion of the purine molecule. Synthesis of the six-membered portion of the purine ring then continues via a carboxylation reaction followed by a complex reaction in which the N-1 atom of the purine ring is introduced utilising aspartic acid and involves elimination of the carbon skeleton of the aspartate moiety as fumarate from the intermediate 5-aminoimidazole-4-N-succinocarboxamide ribonucleotide to produce 5-aminoimidazole 4-carboxamide ribonucleotide. An ATP is consumed in the initial step whereby the amino group of aspartate reacts with the carboxyl group of 5-aminoimidazole-4-carboxylate ribonucleotide. The final two steps in purine biosynthesis involve a formylation reaction utilising 10-formyltetrahydrofolate and a dehydration reaction resulting in ring closure to produce the complete parent ribonucleotide inosine-5'-monophosphate. The actual purine base portion of this ribonucleotide is hypoxanthine.

Inosinate is the precursor of AMP and GMP (Fig. 4.2). AMP synthesis requires the introduction of an amino group at C-6 in place of the carbonyl group, aspartate being the amino donor in a reaction which involves addition of the amino acid to form

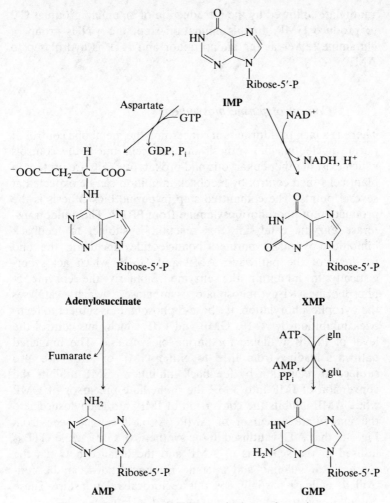

Fig. 4.2 Conversion of IMP to adenine and guanine
ribonucleotides

adenylosuccinate followed by elimination of fumarate (made up
of the C-skeleton of aspartate). Adenylate synthesis involves the
hydrolysis of GTP in the initial step and the second step (fumarate
removal from adenylosuccinate) is catalysed by the same enzyme
which catalyses fumarate removal from 5-aminoimidazole-4-N-
succinocarboxamide ribonucleotide in the formation of IMP.
GMP is synthesised from inosinate via an NAD-dependent oxidation reaction which introduces a carbonyl group at C-2 to produce

xanthylate followed by the introduction of an amino group at C-2 to produce GMP. In this amination step, the γ-NH_2 group of glutamine serves as the amino donor and ATP is hydrolysed to AMP.

Control of purine biosynthesis

There is a lack of information concerning the metabolic control of purine nucleotide biosynthesis in plant cells but if the controls which exist in other eukaryotic and prokaryotic cells are present in plant cells then control by feedback inhibition can be expected at several points. The committed step in purine biosynthesis is the production of phosphoribosylamine from PRPP. The amidotransferase enzyme catalysing this reaction is subject to feedback inhibition by many purine ribonucleotides including the end products of the pathways, AMP and GMP, which act synergistically in inhibiting the enzyme. Similarly the enzyme 5-phosphoribosyl-1-pyrophosphate synthetase, which catalyses the pyrophosphorylation of ribose-5-phosphate, is subject to feedback inhibition by AMP, GMP and IMP which thus control the level of PRPP available for purine biosynthesis. The branched pathways leading from IMP to either GMP or AMP are also subject to regulation by feedback inhibition. GMP inhibits the conversion of IMP into XMP the immediate precursor of GMP while AMP inhibits the conversion of IMP into adenylosuccinate the immediate precursor of AMP. Also, it can be seen from Fig. 4.2 that ATP is utilised in the synthesis of GMP while GTP is utilised in the synthesis of AMP and the balance between the synthesis of guanine and adenine ribonucleotides can be controlled, at least partially, by this reciprocal relationship. These potential control points are summarised in Fig. 4.3.

Biosynthesis of pyrimidines

In contrast to purine biosynthesis where the purine ring is assembled while linked to ribose-5′-phosphate (i.e. via nucleotide intermediates), the pyrimidine ring is synthesised first and then joined to ribose-5′-phosphate to form a nucleotide. Pyrimidine biosynthesis begins with the formation of carbamoyl phosphate, an activated carbamoyl donor, utilising glutamine, bicarbonate and ATP in a reaction catalysed by the enzyme carbamoyl phosphate

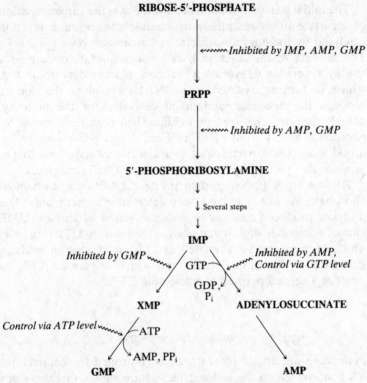

Fig. 4.3 Control points in purine biosynthesis

synthetase:

$$\text{Glutamine} + HCO_3^- + 2ATP \longrightarrow$$

$$\underset{\substack{|\\O^-}}{\overset{\substack{O\quad\quad O\\\|\quad\quad\|}}{H_2N-C-O-P-O^-}} + \text{Glutamate} + 2ADP + P_i$$

Carbamoyl phosphate

Carbamoyl phosphate and aspartate are the precursors used in the synthesis of the pyrimidine ring and the origins of the C and N atoms of the ring are as shown:

C-2, N-3 from $\left\{\begin{matrix} & C & \\ N_3 & {}^4\;{}_5 & C \\ C^2 & {}_1\;{}^6 & C \\ & N & \end{matrix}\right\}$ N-1, C-4, 5, 6
carbamoyl   from aspartate
phosphate

The initial step in pyrimidine biosynthesis is the carbamoylation of aspartate in an essentially irreversible condensation reaction catalysed by the enzyme aspartate transcarbamoylase (Fig. 4.4). The product of the reaction N-carbamoylaspartate undergoes a readily reversible dehydration reaction to form dihydro-orotate which in turn is oxidised in an NAD-dependent reaction to produce the first true pyrimidine derivative of the pathway, namely orotate. This free pyrimidine then acquires a ribose-5'-phosphate moiety from PRPP to form a pyrimidine nucleotide, orotidylate, which participates in a decarboxylation reaction' to produce the 'parent' pyrimidine nucleotide UMP (uridylate).

All the enzymes required to produce UMP from carbamoyl phosphate and aspartate have been detected in higher plants. The uridylate produced can now be phosphorylated by kinases (UMP kinase, nucleoside diphosphate kinase) to produce UTP (Fig. 4.4) which in turn can be aminated at C-6, in a reaction utilising glutamine as the amino donor and which consumes ATP, to produce a second pyrimidine nucleotide CTP.

Control of pyrimidine biosynthesis

Pyrimidine nucleotide biosynthesis is controlled by feedback inhibition exerted on the two enzymes catalysing the first two steps in the biosynthetic pathway, i.e. carbamoyl phosphate synthetase producing carbamoyl phosphate (Fig. 4.4). Both enzymes are inhibited by a variety of nucleotides of which the most potent inhibitor is UMP. However, while aspartate transcarbamoylase is inhibited by the pyrimidine nucleotides UMP, CMP, UDP and UTP but not by purine nucleotides, the glutamine-dependent carbamoyl phosphate synthetase from pea seedlings is inhibited not only by UMP, UDP and UTP, but also by adenine and guanine nucleotides.

Biosynthesis of deoxyribonucleotides

In the absence of any detailed studies on deoxyribonucleotide biosynthesis in plants the pathways involved in their synthesis and the control mechanisms operating cannot be described with any degree of certainty. However, it is possible to draw analogies with other eukaryotic and bacterial biosynthetic pathways and it is known that the level of free deoxyribonucleotides in plant cells is low, indicating that their synthesis is controlled so that it coincides with the onset of DNA synthesis prior to cell division.

The reduction of the ribose moiety to 2-deoxyribose in deoxyribonucleotide biosynthesis occurs at the nucleotide stage while the ribose phosphate is attached to the purine or pyrimidine base. The enzyme catalysing this reaction, ribonucleotide reductase, must be present in plant tissues and has been well studied in mammalian and bacterial systems. The reaction catalysed by ribonucleotide reductase is:

$$\text{Ribonucleoside diphosphate} + \text{NADPH} + \text{H}^+ \longrightarrow$$
$$\text{Deoxyribonucleoside diphosphate} + \text{NADP}^+ + \text{H}_2\text{O}$$

If the plant enzyme is similar to the mammalian enzyme then the actual reaction mechanism will be more complex than is indicated by the equation. Carriers of reducing power such as thioredoxin and glutathione plus flavoproteins may be involved in the transfer of electrons from NADPH to the catalytic site of ribonucleotide reductase.

Purine deoxyribonucleotides are synthesised in a straightforward manner with either ADP or GDP serving as substrate for ribonucleotide reductase:

$$\text{ADP} \xrightarrow{\text{Reductase}} \text{dADP} \xrightarrow{\text{Phosphorylation}} \text{dATP}$$
$$\text{GDP} \xrightarrow{\text{Reductase}} \text{dGDP} \xrightarrow{\text{Phosphorylation}} \text{dGTP}$$

However, in eukaryotes the pyrimidine deoxyribonucleotides dUMP and TMP appear to arise chiefly by hydrolytic deamination of deoxycytidine nucleotides in the manner described in Fig. 4.5, but no studies have been performed on thymidylate synthesis in plant tissues to substantiate this proposed biosynthetic pathway.

The production of TMP from dUMP catalysed by thymidylate synthetase is an interesting reaction in that the methylene tetrahydrofolate used in the reaction serves not only as a one-carbon donor but also as a source of electrons for the reaction. The reason for this is that the methyl group inserted into dUMP is more reduced than the methylene group in the tetrahydrofolate derivative and the two electrons used in the reduction reaction are supplied by the tetrahydrofolate molecule in the form of a hydride

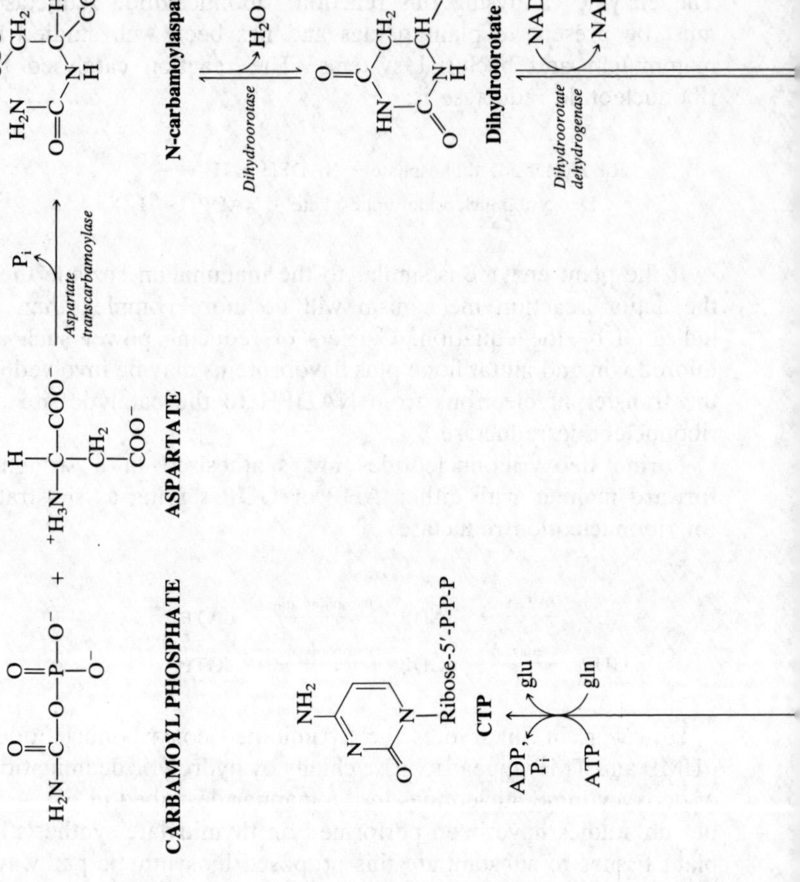

CARBAMOYL PHOSPHATE

+

ASPARTATE

Aspartate transcarbamoylase → P_i

N-carbamoylaspartate

Dihydroorotase → H_2O

Dihydroorotate

Dihydroorotate dehydrogenase : NAD^+ → $NADH, H^+$

CTP

Ribose-5'-P-P-P

gln, ATP → ADP, P_i, glu

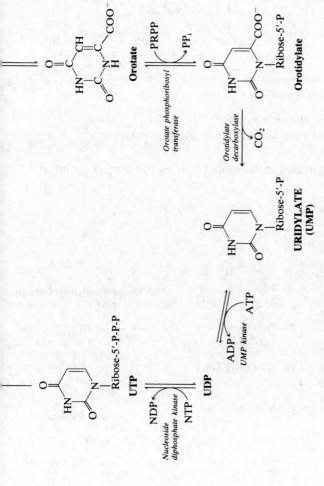

Fig. 4.4 Biosynthesis of the pyrimidine ring

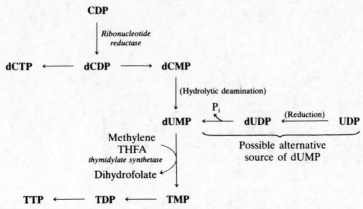

Fig. 4.5 Pyrimidine deoxyribonucleotide biosynthesis

$(H-)$ ion, this hydrogen becoming part of the methyl group of TMP:

dUMP

TMP

Thymidylate synthetase

In the reaction the tetrahydrofolate derivative is oxidised to dihydrofolate and since one-carbon transfers occur at the level of tetrahydrofolate rather than dihydrofolate it is necessary for the cell to regenerate tetrahydrofolate if the reaction is to continue. The regeneration is accomplished by the enzyme dihydrofolate reductase which utilises NADPH as reductant:

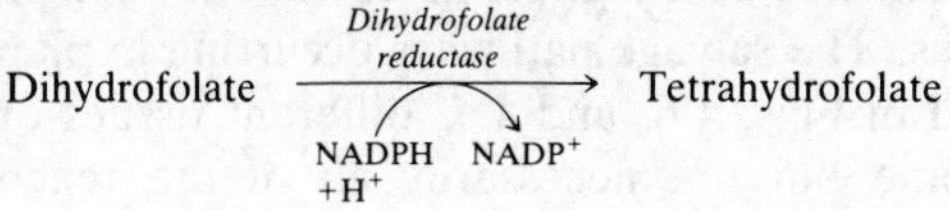

Dihydrofolate $\xrightarrow[\substack{\text{NADPH} \\ +\text{H}^+}]{\substack{\textit{Dihydrofolate} \\ \textit{reductase}}}$ Tetrahydrofolate

NADP$^+$

Control of deoxyribonucleotide biosynthesis

Ribonucleotide reductase is regulated by a set of feedback mechanisms whose function is to ensure that the precursors for DNA biosynthesis, the deoxyribonucleotide triphosphates, are synthesised only in the amounts required for DNA synthesis. Basically, this control is achieved by an excess of one deoxyribonucleotide inhibiting the reduction of all ribonucleoside diphosphates by ribonucleotide reductase. The bacterial enzyme is considered to have two allosteric sites, one controlling the overall activity of the enzyme while the second site regulates substrate specificity. If the ribonucleotide reductase of plants is regulated in the same way as the bacterial enzyme then binding of dATP to the first site would signify an abundance of deoxyribonucleoside triphosphates in the cell and so diminish the overall catalytic activity of the enzyme. ATP binding to this site would reverse this feedback inhibition. Binding of dATP or ATP to the second site would lead to an enhanced reduction of the pyrimidine nucleotides CDP and UDP while TTP binding would promote the reduction of GDP and inhibit the further reduction of pyrimidine ribonucleoside diphosphates. An increase in the level of dGDP and subsequent binding of dGDP to this second allosteric site would stimulate ADP reduction. In this complex way a regulatory mechanism which provides an appropriate supply of all the four deoxyribonucleoside triphosphates required for DNA synthesis can operate within the cell.

Nucleotide biosynthesis by salvage reactions

The turnover of nucleic acids and nucleotides which occurs continuously in cells during normal cell metabolism produces free purine and pyrimidine bases or nucleosides. Nucleotides can be synthesised from these preformed bases by means of salvage reactions which have the advantage over the normal pathways of *de novo* purine and pyrimidine biosynthesis in that they are simpler and much less costly in terms of expenditure of cellular energy resources. The salvage pathways occurring in plant tissues are summarised in Figs. 4.6. and 4.7, different tissues or species undertaking some but not necessarily all of the reactions described. As can be seen, interconversions can occur at the nucleotide, nucleoside and free base levels, the evidence for some of the reactions being more substantial than others. In general it can be seen that free purine and pyrimidine bases can react with PRPP to give mononucleotides and PP_i; thereby giving a route for the reutilisation of the free bases produced in nucleic acid degradation. However, free thymine does not appear to be reused although the nucleoside thymidine can be rapidly phosphorylated by kinases to TTP. In the purine salvage pathways guanine derivatives appear to be less extensively metabolised than the corresponding adenine derivatives.

There is evidence for the existence of two independent pools of IMP for purine nucleotide biosynthesis in several plant tissues, including soybean embryonic axes, germinating wheat embryos and tea shoot tips. One of these pools arises from *de novo* synthesis while the other pool arises from salvage pathways. There appears to be little or no cross-over between these pools and their sizes are subject to independent regulation.

Degradation of purines

Aerobic catabolism of purines occurs primarily in senescing tissue and adenine and guanine share a common breakdown pathway from the level of xanthine onwards. This pathway is illustrated in Fig. 4.8 and shows that adenine is first deaminated to hypoxanthine prior to oxidation to xanthine while guanine degradation involves only a deamination step to produce xanthine. Further degradative steps produce uric acid, allantoin and allantoic acid, this latter compound functioning as an important nitrogenous storage compound in many plant tissues. The enzymes uricase and allantoinase, which are involved in the production of allantoin and allantoic acid respectively, are found localised in the cell in peroxisomes and glyoxysomes. The allantoic acid produced during the breakdown of purines can be further metabolised to urea and glyoxylate if the enzyme allantoinase is present in tissues. Coffee and tea plants contain significant amounts of caffeine and related

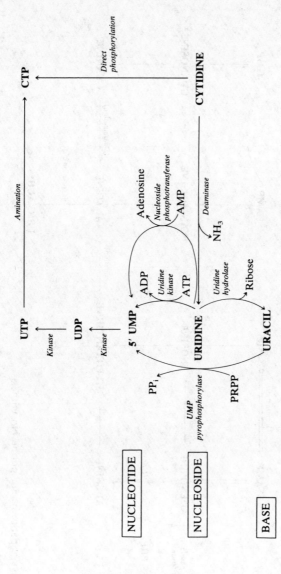

Fig. 4.6 Salvage pathways for pyrimidine biosynthesis

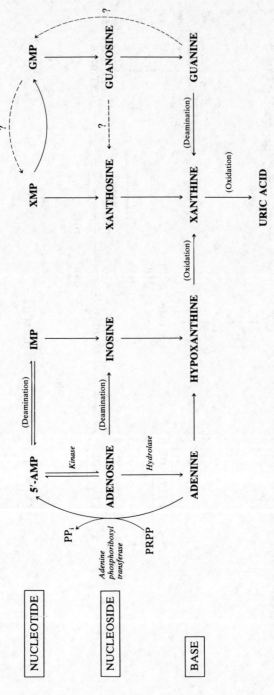

Fig. 4.7 Salvage pathways for purine biosynthesis

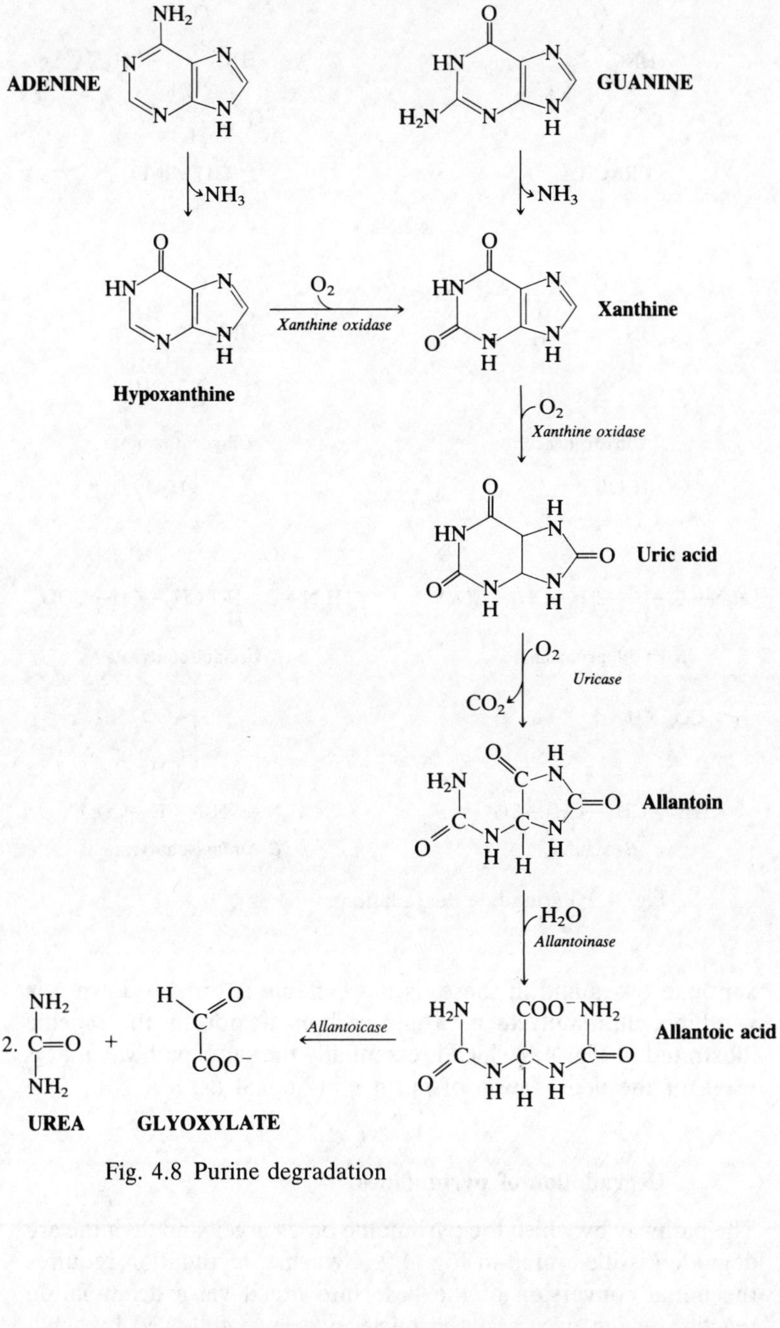

Fig. 4.8 Purine degradation

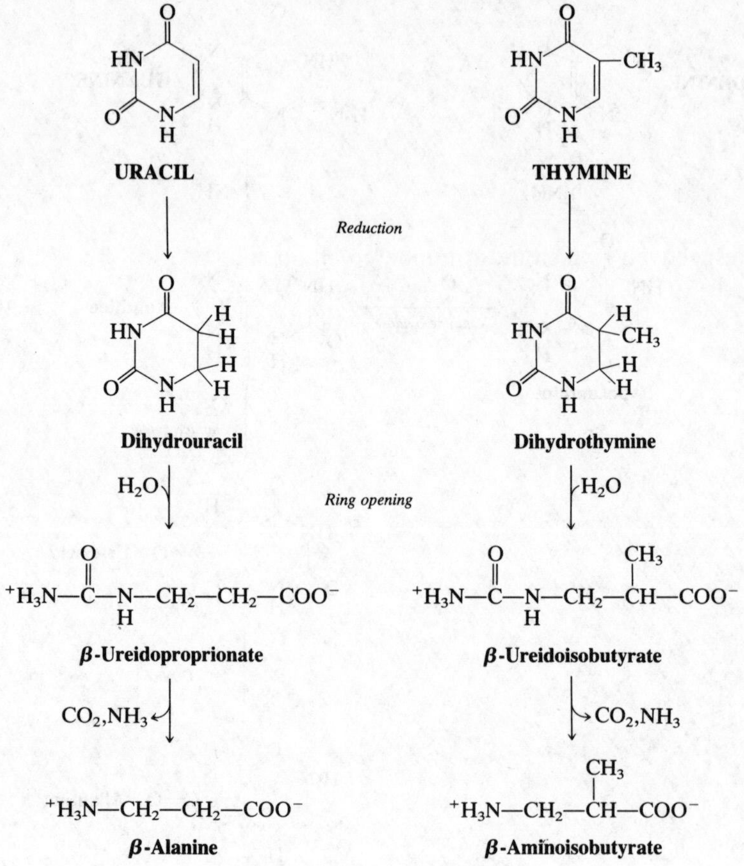

Fig. 4.9 Pyrimidine degradation

xanthine bases and in these tissues caffeine is broken down, via xanthine, ultimately to urea and carbon dioxide by the scheme illustrated in Fig. 4.8 which is essentially the same pathway that is used for the degradation of purines in animal tissues.

Degradation of pyrimidines

The pathway by which the pyrimidine bases uracil and thymine are degraded is illustrated in Fig. 4.9. Cytosine degradation requires the initial conversion of the base into uracil via a deamination reaction which occurs at the nucleoside level followed by a hydrolysis of the uridine produced to form the free base uracil:

Cytidine $\xrightarrow[-NH_2]{\text{deamination}}$ Uridine $\xrightarrow[\text{Ribose}]{\text{hydrolysis}}$ Uracil $\longrightarrow$

Degradation to β-alanine

Uracil and thymine degradation involve the same sequence of reactions beginning with reduction of the free bases to the dihydropyrimidine derivatives dihydrouracil and dihydrothymine respectively, which in turn undergo an opening of the pyrimidine ring to produce the β-ureido-amino acids, β-ureidoproprionate and β-ureidoisobutyrate. The subsequent elimination of CO_2 and ammonia from these acids results in the formation of β-alanine and β-aminoisobutyrate which can ultimately be metabolised by the oxidative pathways of carbon metabolism in the cell or, in the case of β-alanine be used in the biosynthesis of coenzyme A.

Cyclic nucleotides in plants

$3',5'$-cyclic AMP (cAMP) is produced by the action of the enzyme adenylate cyclase on ATP and degraded by the action of a specific phosphodiesterase which hydrolyses it to $5'$-AMP:

(i) Synthesis

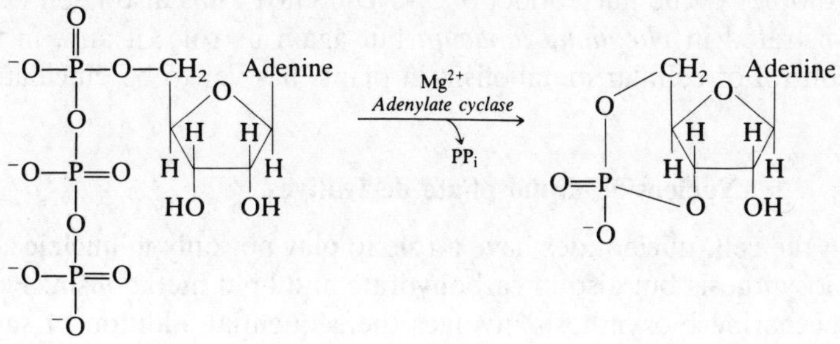

ATP

cAMP

(ii) Degradation

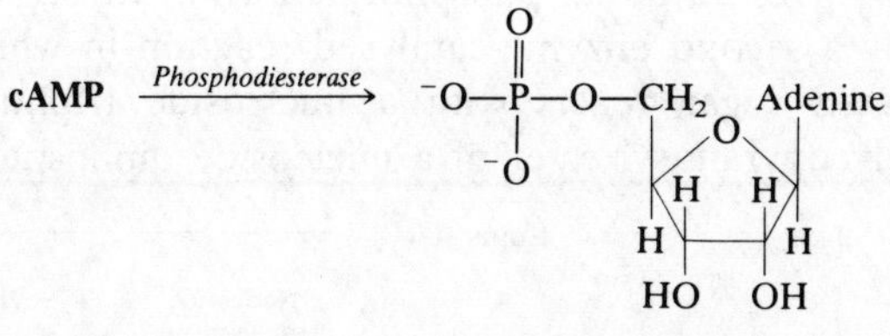

5'-AMP

The presence of cyclic AMP and its role as a second messenger in hormone action in animal cells has been well established over the last 20 years but it is only recently that the presence of the enzyme system responsible for the metabolism of cAMP, i.e. adenylate cyclase and the specific phosphodiesterase, has been demonstrated in plant cells, although the existence of cAMP in plant cells has been known for a considerable time. The presence of these two key enzymes of cAMP metabolism has been demonstrated in tissues of *Phaseolus vulgaris* and other plant species.

As a result, the biochemical potential now exists in plant cells for either a secondary messenger role for cAMP analogous to that in animal tissues or a primary messenger role similar to that in certain bacteria. In both the primary messenger system of microorganisms and the secondary messenger system of animal cells there is a central role for a protein which specifically binds cAMP and subsequently initiates a train of metabolic events. Such cAMP-binding proteins have been demonstrated in *Phaseolus*, *Hordeum* and other higher plants and as some of the effects of gibberellic acid on plant development can be mimicked by cAMP there has been speculation that gibberellic acid responses may involve a second messenger. However, a definite role for cAMP in cellular metabolism in plant tissues has yet to be demonstrated. Another cyclic nucleotide, 3′,5′-cyclic GMP, has also been demonstrated in *Nicotiana tabacum* but again its role, if any, in the control of cellular metabolism in plants has yet to be elucidated.

Nucleoside diphosphate derivatives

In the cell, nucleotides have a role to play not only in nucleic acid biosynthesis but also in carbohydrate and lipid metabolism. Polysaccharide biosynthesis involves the sequential addition of sugar monomer units to the growing polysaccharide chain, the activated form of the sugar monomers used in this biosynthesis being the nucleoside diphosphate sugar derivatives. The activation process involves an initial phosphorylation of the sugar residue followed by a second enzyme-catalysed reaction in which the phosphorylated sugar reacts with a nucleoside triphosphate to form a glycopyranosyl ester of a nucleoside diphosphate:

Sugar $\longrightarrow$ Sugar-1-$\textcircled{P}$ $\longrightarrow$ NDP-sugar

ATP ADP

Nucleoside $PP_i \rightarrow 2P_i$ (sugar nucleotide)
triphosphate
(NTP)

The inorganic pyrophosphate is hydrolysed by pyrophosphatase, thus the production of the sugar nucleotide involves the hydrolysis of two high energy phosphate linkages. The sugar nucleotide is now able to act as a donor of activated glycosyl groups for polymerisation reactions and some of these reactions, summarising the possible fates of the most abundant nucleoside diphosphate sugar in plants, UDP-glucose, are illustrated in Fig. 4.10. The structure of UDP-glucose is illustrated separately:

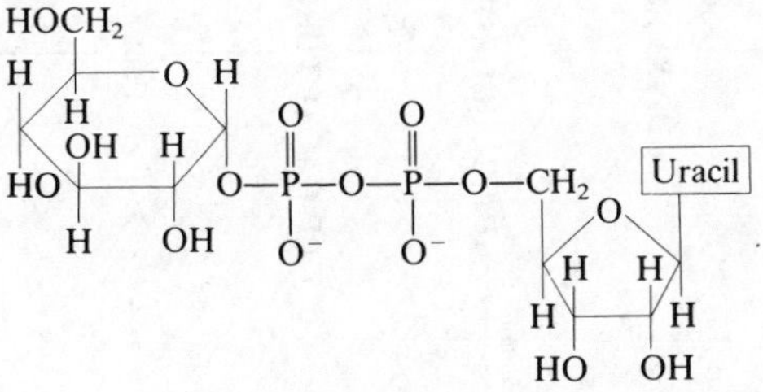

UDP-glucose

UDP-glucose is involved in the synthesis of galactinol, the galactosyl donor in the synthesis of stachyose and verbascose formed by the transfer of additional α-D-galactosyl units on to the 6-hydroxyl group of the galactose units in raffinose which is itself a storage form of galactosyl, glucosyl and fructosyl residues. Stachyose and its homologues appear to act as 'anti-freeze' agents in plants. UDP-glucose also provides the precursors for plant cell wall biosynthesis and serves as a precursor for the biosynthesis of starch, a storage polysaccharide, which serves as a rapidly metabolisable food reserve in cells. The most common food sugar, sucrose, is formed in all green plants and nowhere else utilising UDP-glucose and fructose-6-phosphate:

$$\text{UDP-glucose + Fructose-6-P} \xrightarrow{\substack{\textit{Sucrose phosphate} \\ \textit{synthase}}} \text{UDP + Sucrose-6-}\textcircled{P}$$

$$\text{Sucrose-6-P} \xrightarrow{\substack{\textit{Sucrose phosphate} \\ \textit{synthase}}} \text{Sucrose + P}_i$$

SYNTHESIS

$$\text{Sucrose} \xrightarrow{\textit{Invertase}} \text{Glucose + Fructose} \quad \text{DEGRADATION}$$

Sucrose is extremely water soluble and chemically inert (except to acid hydrolysis) and serves in plants principally as a transport sugar. This disaccharide form of sugar transport is advantageous to the plant since the disaccharide has a higher osmotic potential than

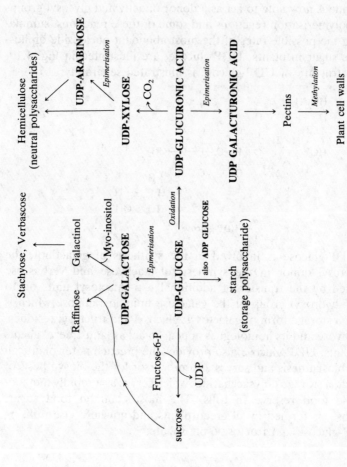

Fig. 4.10 Possible fates of some sugar nucleotides

the equivalent amount of carbon transported as a monosaccharide. The nature of the nucleotide handle of the sugar nucleotide can be varied. ADP-glucose has been implicated in starch biosynthesis catalysed by starch synthetase while GDP-glucose may also function as a glucose donor in cell wall biosynthesis.

Derivatives of CTP, in contrast to those of UTP, GTP and ATP, function principally not in carbohydrate metabolism, but in lipid biosynthesis. Although little is known about the details of their involvement in plant cells, two cytidine derivatives have been implicated:

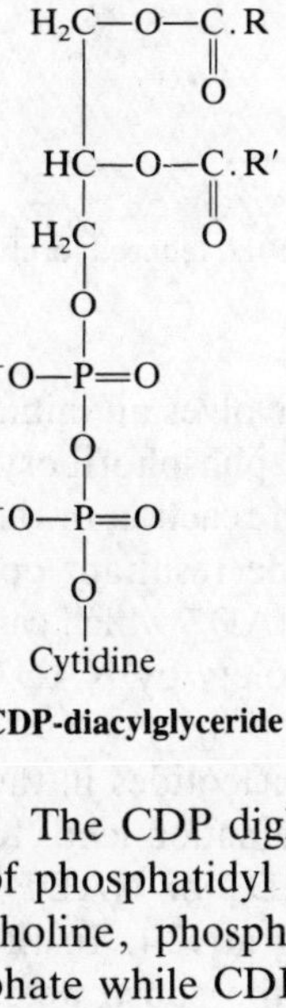

CDP-diacylglyceride 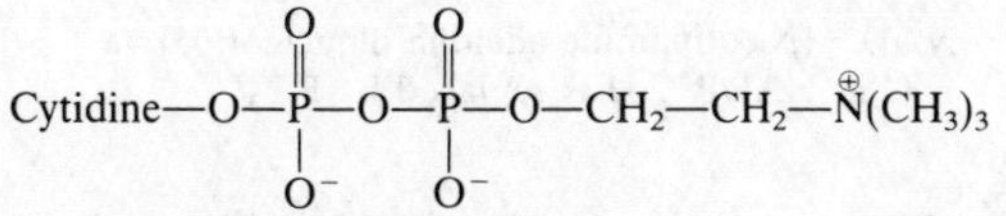   **CDP-choline**

The CDP diglyceride produced is involved in the biosynthesis of phosphatidyl serine, phosphatidyl ethanolamine, phosphatidyl choline, phosphatidyl inositol and phosphatidyl glycerol phosphate while CDP-choline is involved in the synthesis of phosphatidyl choline by an alternative pathway.

Pyridine nucleotides

Pyridine nucleotides are dinucleotides derived from the pyridine nicotinamide and the purine adenine and are among the principal components of the acid-soluble nucleotides extracted from tissues of higher plants. The main role of these nucleotides in cellular metabolism is in the oxidation – reduction reactions of the cell. The structure of NAD^+ and $NADP^+$ is illustrated below while the biosynthesis of these molecules is outlined in Fig. 4.11.

NAD$^+$ (Nicotinamide adenine dinucleotide)   NAD(P)H, reduced form
(* in NADP$^+$, H is replaced by PO$_3^=$)

The synthesis of these nucleotide derivatives involves an initial condensation reaction between nicotinate and 5′-phosphoribosyl pyrophosphate followed by a second condensation reaction of the product nicotinate ribonucleotide, with ATP. The resultant de-samido-NAD$^+$ undergoes amination to produce NAD$^+$ which can be phosphorylated at the 2′-OH of the AMP moiety by NAD$^+$ kinase to produce NADP$^+$.

The ratios of oxidised and reduced pyridine nucleotides in the cell vary with environmental conditions, e.g. illumination leads to increased NADPH levels. Reduction of the NAD$^+$ or NADP$^+$ molecule involves reduction of the pyridine ring at C-4. In the oxidation of substrate, one hydrogen atom of the substrate is directly transferred to NAD$^+$ while the other appears in the solvent, however, both electrons lost by the substrate are transferred to the nicotinamide ring:

$$NAD + Substrate\text{-}H_2 \rightleftharpoons NADH + Substrate \text{ (oxidised)} + H^+$$

The reduction of the pyridine ring of NAD$^+$ results in the appearance of strong absorption properties at 340 nm which can be used to monitor the appearance or disappearance of NADH or NADPH during oxidation-reduction reactions. In general, it seems that NADPH is used almost exclusively in reductive biosyn-

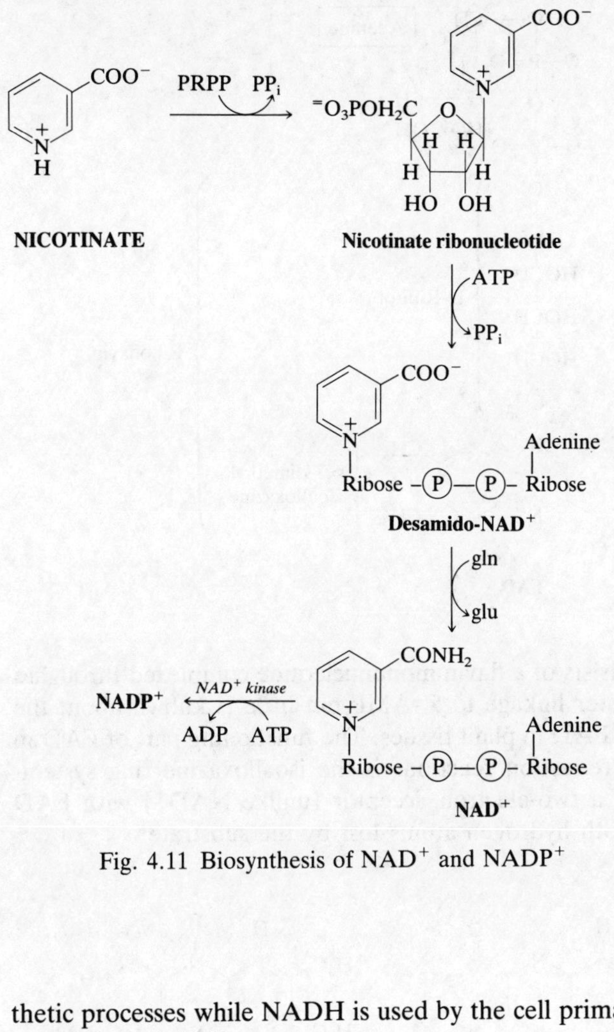

NICOTINATE   **Nicotinate ribonucleotide**

Desamido-NAD$^+$

NAD$^+$

Fig. 4.11 Biosynthesis of NAD$^+$ and NADP$^+$

thetic processes while NADH is used by the cell primarily for the generation of ATP.

Flavin nucleotides

The other major electron carrier in the oxidation reactions of the cell is flavin adenine dinucleotide, abbreviate to FAD (oxidised form) and FADH$_2$ (reduced form). Flavin coenzymes are usually tightly bound to proteins and cycle between the reduced and oxidised states while still bound to the same protein molecule. See the structure of FAD illustrated.

FAD

FAD consists of a flavin mononucleotide complexed through a phosphodiester linkage to 5'-AMP but little is known about the synthesis of FAD in plant tissues. The functioning part of FAD in oxidation – reduction reactions is the isoalloxazine ring system, FAD being a two-electron acceptor (unlike NAD^+) with FAD accepting both hydrogen atoms lost by the substrate:

Oxidised form (FAD)   Reduced form ($FADH_2$)

Reduction of FAD results in a bleaching of the characteristic yellow colouration of oxidised flavins and this allows the reduction process to be monitored spectrophotometrically by following the decrease in absorption of the flavin nucleotide at 375 nm and 450 nm as the reduction process proceeds.

Suggestions for further reading

General

Adams, R. L. P., Burdon, R. H., Campbell, A. M., Leader, D. P. & Smellie, R. M. S. (1981) *The Biochemistry of Nucleic Acids* (9th edn). Chapman & Hall: London & New York.

Metzler, D. E. (1977) *Biochemistry. The Chemical Reactions of Living Cells.* Academic Press: N.Y., San Francisco & London.

Stryer, L. (1981) *Biochemistry* (2nd edn). W. H. Freeman: San Francisco.

Specific

Purine nucleotides

Anderson, J. D. (1979) *Plant Physiol.*, **63**, 100–4.

Cyclic AMP

Amrheim, N. (1977) *Ann. Rev Plant Physiol.*, **28**, 123–32.

Brown, E.G., Edwards, M. J., Newton, R. R. & Smith, C. J. (1980) *Phytochem.*, **19**, 23–30.

Brown, E. G., Newton, R. P. & Smith, C. J. (1980) *Phytochem.*, **19**, 2263–6.

Nucleotide sugars

Rodaway, S. & Marcus, A (1979) *Plant Physiol.*, **64**, 975–81.

5

Nucleic acids

Deoxyribonucleic acid

With the exception of certain viruses, the biological macromolecule carrying the genetic information in the living cell is the deoxyribonucleic acid (DNA) molecule. The genetic information is contained within the unique linear sequence of deoxyribonucleotide units which comprise the DNA molecule of the particular organism and the genes responsible for the particular characteristics of the organism are arranged in a linear fashion along the DNA molecule. Expression of the genetic information encoded within the genes involves primarily the production of a messenger ribonucleic acid molecule (mRNA) in which the sequence of bases reflects accurately the sequence of bases in the gene from which it has been transcribed. The base sequence of the mRNA molecules is then translated by the protein-synthesising machinery of the cell into the unique linear sequence of amino acids which goes to make up a cellular protein.

Intracellular location of DNA

In the eukaryotic cell the DNA is localised mainly within the nucleus where it exists complexed with proteins and is termed chromatin. The individual aliquots of chromatin in the nucleus become visible as chromosomes during mitosis and meiosis. Much smaller amounts of DNA are also to be found in mitochondria and chloroplasts. The amount of DNA per somatic cell nucleus is the same for all types of cell within an organism but the DNA content of the gamete nuclei is exactly half that of the somatic cell nuclei, i.e. the DNA content of the nucleus is directly related to the chromosome content of the nucleus depending on whether the cell is diploid or haploid. Also, the overall base composition of DNA is constant for all cells within a given species but the base composition does vary from species to species.

Composition and structure of DNA

The common monomeric units from which DNA is synthesised are the four deoxyribonucleotides containing the bases adenine (A), guanine (G), cytosine (C) and thymine (T) although many DNA molecules also contain small yet significant amounts of other bases, e.g. 5-methyl cytosine. There is a wide variation in the molar proportions of bases in the DNA molecules of different species as can be seen in Table 5.1, yet certain regularities can be detected in the chemical composition of DNA and these are summarised below:

    1 Σpurine bases = Σpyrimidine bases;

    2 Σamino bases (A+C) = Σketo bases (G+T);

    3 A and T are present in equimolar amounts;

    4 G and C are present in equimolar amounts.

The equivalence of A and T and of G and C is commonly referred to as Chargaff's rule (it was Chargaff who first drew attention to these regularities) and this observation proved to be of great importance in relation to the structure of DNA. However, from Table 5.1 it can be seen that in wheat germ and *Daucus carota* in particular, guanine and cytosine are not present in equimolar amounts. This apparent discrepancy is explained by the deficiency in cytosine being compensated by the presence of the modified base 5-methyl cytosine.

Table 5.1 Base composition of DNA from various plant sources

Source of DNA	Molar proportions of bases (moles nitrogenous constituents/100 g-atoms P)				
	Adenine	Guanine	Cytosine	Thymine	5-Methyl cytosine
Wheat germ (*Triticum sp.*)	27.3	22.7	16.8	27.1	6.0
Daucus carota	26.7	23.2	17.3	26.8	6.0
Cucurbita pepo	30.2	21.0	16.1	29.0	3.7
Allium cepa	31.8	18.4	12.8	31.3	5.4

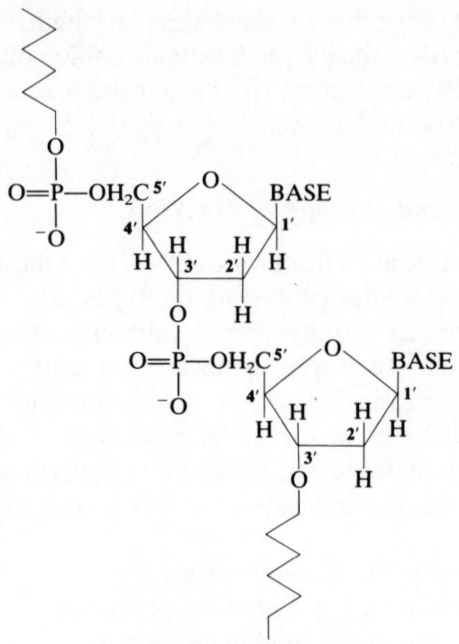

Fig. 5.1 Phosphodiester linkage between two deoxynucleoside
monophosphate residues in the DNA molecule

Within the primary structure of the DNA molecule, the deoxynucleoside monophosphate residues are linked together by phosphodiester bonds between C-3′ of one deoxyribose and C-5′ of the
next deoxyribose (Fig. 5.1).

The DNA molecule consists of two of these polynucleotide
chains, each chain being a right-handed helix with a pitch of
3.4 nm having 10 base pairs per turn of the helix and the base
pairs being spaced at intervals of 0.34 nm (Fig. 5.2).

The two chains of the helix are interlocked and twisted about
the same axis with the phosphate groups on the outside of the helix
forming a double sugar-phosphate backbone and the purine and
pyrimidine bases on the inside of the helix with their planes at 90°,
i.e. perpendicular, to the long axis of the helix. The two polynucleotide chains run in opposite 'chemical' directions (Fig. 5.3), i.e.
they are antiparallel, and the two chains are held together by
hydrogen bonding which is base specific, occurring only between
A and T or G and C bases which is consistent with Chargaff's rule.

Two hydrogen bonds are formed between an adenine–thymine
base pair and three hydrogen bonds between a guanine–cytosine
base pair (Fig. 5.4). From this model for DNA structure originally

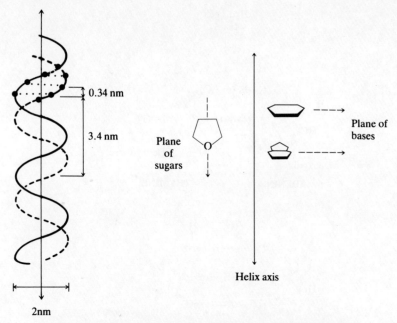

Fig. 5.2 Diagrammatic representation of the double helical
structure of DNA and orientation of the planes of the
sugar residues and bases

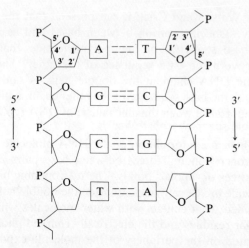

Fig. 5.3 Diagrammatic representation of part of a
polynucleotide chain in DNA

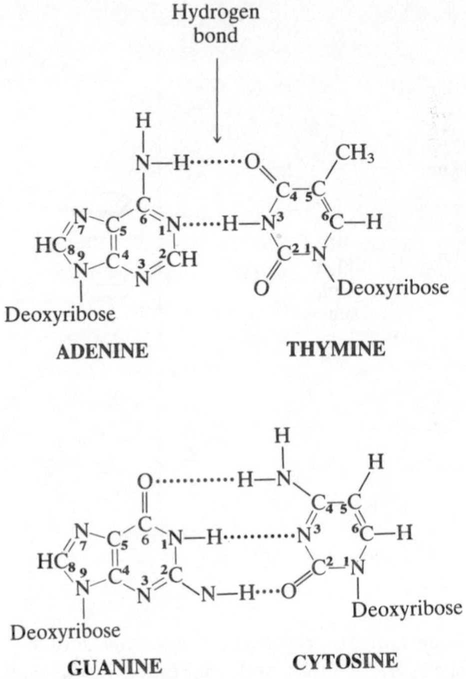

Fig. 5.4 Base pairing of adenine and thymine and guanine and cytosine

put forward by Watson and Crick it can be seen that the DNA molecule involves specific pairing between bases and as a consequence the base sequence of one polynucleotide chain will automatically govern the base sequence of the other. The base sequence of one DNA chain in the double helix is said to be complementary to the base sequence of the other chain, a factor of considerable importance when the mechanism of DNA replication is considered.

The double-helical conformation of the DNA molecule is stabilised by hydrogen-bonding forces between base pairs and by base stacking forces or 'apolar' bonds. The hydrophobic bases of the DNA molecule are closely stacked inside the double helix and effectively shielded from contact with water molecules while the hydrophilic sugar residues and the electrically charged phosphate groups are located on the periphery of the molecule exposed to water. The hydrophobic nature of the bases makes contact with the aqueous medium entropically unfavourable and the vertical stacking of bases within the double-stranded DNA molecule thus helps to minimise contact with the aqueous medium and increases the stability of the double helical structure of the DNA molecule.

Chromatin structure and composition

The molecular weight of a DNA molecule ranges from 10^6 to more than 10^{10} daltons depending upon the organism from which it originates. It is difficult to determine an accurate molecular weight for a molecule of such an enormous size because estimations are complicated by the difficulties experienced in preparation of whole DNA molecules. These large high molecular weight DNA molecules are very susceptible to hydrodynamic shearing forces which cause some degradation of the DNA, e.g. during pipetting, and also to the degradative action of contaminating nucleases in the DNA preparation.

The large amount of DNA in the eukaryotic cell, with the exception of mitochondrial and chloroplast DNA, must be packed into the cell nucleus. This 'packaging' imposes the necessity for supercoiling of the DNA (Fig. 5.5) a process during which the simple double helix of the DNA molecule undergoes some distortion. The DNA molecule is unlikely to assume a stable supercoiled configuration spontaneously and in the nucleus of a eukaryotic cell the DNA is found as a complex with proteins, this complex being termed chromatin.

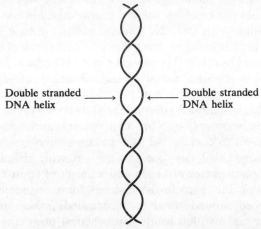

Double stranded DNA helix ⟶   ⟵ Double stranded DNA helix

Fig. 5.5 Section of duplex DNA containing superhelical turns

Table 5.2   Classes of histones found in plant tissues

Class	Description	Alternative name
H1	Very lysine rich	I or f1
H2b	Lysine rich	IIb2 or f2b
H2a	Arginine – lysine rich	IIb1 or f2a2
H3	Arginine rich	III or f3
H4	Glycine – arginine rich	IV or f2a1

Chromatin proteins   There are two classes of protein found in chromatin, histones which are basic proteins rich in the basic amino acids arginine and lysine and the acidic or non-histone chromatin proteins. In terms of mass, the histone proteins are easily the most abundant of these two classes of protein and in a chromatin sample the mass of histones approximately equals the mass of DNA. Chromatin structure is determined largely by the histones which are responsible for the compaction of the DNA within the nucleus of the cell. There are five main classes of histones found in plants (Table 5.2), these histones being present in all chromatin samples examined.

Nucleosome structure   The use of nucleases which hydrolyse specific phosphodiester bonds in the DNA molecule has made it possible to break down chromatin into 'subunits'. However, only the elementary subunit of chromatin structure, the nucleosome, has been identified and characterised in any detail. The nucleosome 'core particle' contains about 146 base pairs of DNA and a protein octamer which consists of two each of the four core histones H2A, H2B, H3 and H4 but no H1 which, although not involved in the structure of the nucleosome, is envisaged as being associated with it. This 'core particle' appears to contain that portion of the nucleosomal DNA most tightly bound to the surface of the histone octamer (Fig. 5.6) and in rye, this nucleosome core has been shown to contain 140 base pairs.

It is envisaged that this core particle is roughly cylindrical in shape having a diameter of 11 nm and a height of approximately 5.5 nm. Within this particle the histones form a core with the DNA wrapped around them in supercoiled fashion to form between one and two left-handed superhelical turns, the superhelix containing about 80 base pairs per turn. The sequences of

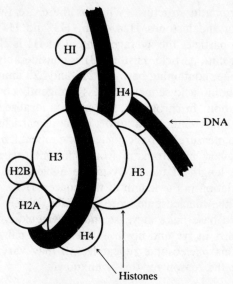

Fig. 5.6 Schematic diagram of chromatin structure indicating possible structure of a nucleosome

amino acid residues of the histones which make up the protein core of the nucleosome 'core particle' exhibit two general features:

1 They show a non-uniform distribution of amino acids within the protein molecule, the C-terminal two-thirds of the molecule being similar in amino acid composition to a globular protein while the N-terminal third is more basic.
2 All of these core proteins, but in particular H3 and H4 (the arginine-rich histones) are highly conserved throughout evolution – histone H4 from pea seedlings differs from that of bovine thymus by only two amino acids. This type of evidence has been used to suggest that histones play a very fundamental role in nucleosome structure and that a very large part of their amino acid sequence is necessary for this role. The nature of the DNA – histone binding is probably via an electrostatic interaction between negatively charged phosphate groups of DNA and the positively charged basic amino acid residues (chiefly arginine and lysine) in histones.

The nucleosome core particle plus the 'spacer' or 'linker' DNA which joins the nucleosome core particles plus lysine-rich histone H1 are the components which make up the mononucleosome, the next level of chromatin structure. Whereas there are two molecules of each of the histones H2a, H2b, H3 and H4 in the nucleosome core particle, the average content of H1 is closer to one molecule per core particle. Histone H1 is considerably larger than a core histone, containing between 210 and 220 amino acid residues and its amino acid sequence is less stringently conserved throughout evolution. In chromatin, H1 appears localised at the point where the DNA enters and leaves the core particle.

It is possible to measure the basic nucleosome repeat length and this is found to vary in chromatin from different tissues. The variability in the length of the DNA in the nucleosome repeat arises from a variation in the length of the 'spacer' DNA regions which separate the nucleosome core particles, these lengths varying from zero base pairs in yeast and lower eukaryotes, to about 60 base pairs in rye and up to about 80 base pairs in sea urchin sperm. This nucleosome repeat length may vary during development but the reason for this is unknown.

The arrangement of nucleosomes in chromatin   The nucleosomes must pack in some regular way into the higher ordered structures which have been observed under the electron microscope. One of these structures, the 10 nm fibril, appears to consist of a linear array of nucleosomes which is in turn supercoiled to form a 30 nm fibre. The nucleosomes within the 10 nm fibril are arranged edgewise rather than with their faces touching and the 30 nm fibre consists of a continuous superhelix of nucleosomes with between six and seven nucleosomes per turn. This 30 nm fibre must then be compacted by another two orders of magnitude, presumably using the principle of supercoiling, to make a mitotic chromosome, but information at this level of chromosome organisation is lacking. However, it does appear that the long continuous DNA of interphase chromosomes is organised in looped domains which contain approximately 80 kilobase pairs and are most probably held together by RNA and protein.

There is evidence in the case of transcriptionally active chromatin, i.e. those genes actively being transcribed into RNA, that the nucleosome structures are possibly modified in composition and structure compared with the 'usual' nucleosome described previously, but regulation of transcription is not likely to be determined simply by the presence or absence of nucleosome core histones. In the nucleus of the cell there are 15–20 major non-histone nuclear proteins (acidic proteins) ranging in molecular weight from 10,000 to 200,000 daltons plus many other non-histone proteins present in much smaller amount. It is an interesting observation that functionally active genes, although having their normal complement of histones, may have increased levels of non-histone proteins associated with them. The significance of this observation has yet to be explained, but it may be that it is these proteins which have an important role to play in the control of gene expression, while the positively charged histones and protamines are important in the maintenance of chromatin structure.

DNA sequences in eukaryotic chromosomes

The DNA of most eukaryotic cells is composed of bases arranged in either unique sequences, moderately repeated sequences or highly repeated sequences.

Highly repeated DNA sequences A large proportion of the DNA in a plant cell nucleus is in the form of highly repeated sequences, the amount varying in different species. These highly repeated sequences are present in all eukaryotic DNA and can comprise up to about 50 per cent of the genome with an 'average' value of 15 per cent. The sequence of bases repeated occurs in blocks of several million repeat units which often have a G+C content quite different from that of the bulk of cell DNA. The DNA containing these highly repeated sequences has a buoyant density greater than that of bulk cellular DNA and thus bands at a different or 'satellite' position in caesium chloride density gradients. These 'satellite' DNA regions are usually found clustered at the centromere region of chromosomes and could be responsible for the maintenance of the structural integrity of the chromosome. However, apart from this suggestion, no satisfactory biological role has yet been proposed for these highly repeated DNA sequences

Moderately repeated DNA sequences These DNA sequences include those genes coding for histones and ribosomal RNA. The number of copies of ribosomal RNA genes per genome is very variable ranging from 1,580 copies per telophase nucleus in artichoke (*Helianthus tuberosus*) to 13,300 in onion (*Allium cepa*). The sequences of the two major species of ribosomal RNA are arranged in tandem along the genome with spacer DNA separating them.

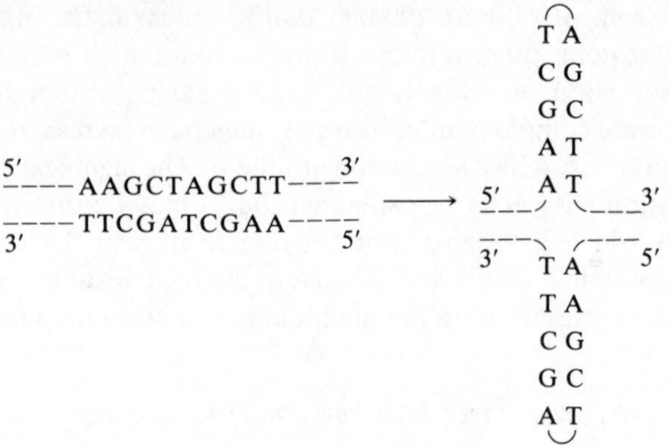

Fig. 5.7 Inverted repeat sequence in DNA

Unique DNA sequences These sequences are more difficult to detect than the repeated DNA sequences and comprise only a small proportion of the genome. Most of these non-repeated sequences appear shorter than 1,000 base pairs in wheat, rye, barley and oats. These unique DNA sequences are thought to code for proteins which have specialised cellular functions and the unique and repeated regions of DNA appear to alternate over a substantial portion of the genome.

Palindromic regions Another structural feature of eukaryotic DNA is the high proportion of regions containing inverted repetition sequences. The two-fold symmetry of these sequences allows the molecule to open up and form a 'cruciform' structure (Fig 5.7), the majority of these sequences present in eukaryotic DNA being between 300 and 1,200 nucleotides long and, in some species, comprising up to 30 per cent of the genome.

Chloroplast DNA

Chloroplasts of several plant species have been shown to contain large circular supercoiled DNA (ctDNA) molecules. The contour length of ctDNA and hence the molecular weight varies with the species (Table 5.3) and the entire chloroplast genome is represented by one circular DNA molecule even though chloroplasts are polyploid and can contain between 20 and 60 copies of ctDNA per chloroplast.

The significance of the size differences between species is uncertain but could suggest that some parts of ctDNA are not

Table 5.3   Molecular weights of chloroplast DNA

Organism	Molecular weight
Zea mays	9.11×10^7
Avena sativa	9.22×10^7
Pisum sativum	9.50×10^7
Spinacia oleracea	10.14×10^7

strictly required for chloroplast function. One interesting observation has been the discovery of covalently inserted ribonucleotides in ctDNA (animal mitochondrial DNA also contains some ribonucleotides). It has been estimated that ctDNA from *Pisum sativum and Spinacia oleracea* contains approximately 18 ribonucleotides while that of *Lactuca sativa* contains approximately 12 ribonucleotides, and although the significance of the presence of such ribonucleotides in ctDNA is uncertain, it has been speculated that they may play a role in ctDNA replication.

Plant mitochondrial DNA

Plant mitochondrial DNA (mtDNA) is a circular molecule with a molecular weight estimated to be in the range $220–1,600 \times 10^6$ daltons. Thus the plant mtDNA is at least seven times the size of the animal mitochondrial genome although there is no evidence that plant mtDNA has any additional informational content. The nucleo-cytoplasmic system of the cell codes for and synthesises the majority (90 per cent) of the proteins and all of the lipids and carbohydrates of the mitochondria. However, the relatively small mitochondrial genome does code for most, if not all, the mitochondrial RNA and up to 10 per cent of the mitochondrial protein. In contrast to nuclear DNA, the DNA from mitochondria and chloroplasts is devoid of histone proteins.

DNA replication

In the eukaryotic chromosome, DNA is complexed with an equal weight of histones to form nucleosome structural units, the periodic repeating unit of chromatin characteristic of all eukaryotic chromosomes. When eukaryotic chromosomes are replicated not only must the DNA molecule be duplicated but also new nucleosomal units must be produced. Also, if gene expression in the daughter cells is to remain unaltered then quiescent and actively transcribed genes, which have different structures within the chromosome, must be accurately duplicated in both base sequence and structure in the daughter cells.

DNA replication in the nucleus of eukaryotic cells occurs simultaneously at many regions along the chromosomes but this process does not appear to utilise unique DNA sequence signals. In contrast, mitochondrial DNA synthesis does utilise specific sequence signals which are recognised by specific proteins. Chloroplast DNA synthesis has been demonstrated in isolated chloroplasts indicating the existence of a chloroplast DNA polymerase but little else is known concerning the mechanism of DNA synthesis in this organelle. Several different types of DNA polymerase enzymes exist in eukaryotic cells, all of them capable of utilising deoxynucleoside-5'-phosphates to extend a growing polynucleotide chain in the 5'- to 3'-direction but these DNA polymerases cannot participate in polynucleotide synthesis unless a primer molecule (polyribonucleotide or polydeoxyribonucleotide) is available.

Enzymes involved in DNA replication   The several different DNA polymerases identified in eukaryotic cells are listed in Table 5.4. DNA polymerases-α and -β have been found in plant tissues along with a DNA polymerase-γ-like enzyme, although in plants the γ-like mitochondrial enzyme differs from that in vertebrates as it also shows some α-like properties. The polymerases differ in the type of polynucleotide product synthesised, DNA polymerase-α synthesising relatively short polynucleotide chains utilising polyribonucleotide primers while DNA polymerase-β adds single nucleotides to an 'activated' DNA primer, a property

Table 5.4   DNA polymerases in eukaryotic cells

Type	Cellular location and function
DNA polymerase-α	Nuclear polymerase. Replication of nuclear DNA
DNA polymerase-β	Nuclear polymerase. Repair of nuclear DNA
DNA polymerase-γ	Found in nucleus and mitochondria. Replication of mitochondrial DNA. Nuclear function uncertain
DNA polymerase-δ	Properties similar to DNA polymerase-α but has associated exonuclease activity

consistent with the role of this enzyme in a repair-type of DNA synthesis.

Mechanism of DNA replication The DNA molecule is replicated semi-conservatively, each of the strands in the DNA molecule acting as a templet for the synthesis of a new complementary strand. This results in the synthesis of two molecules of double-stranded DNA with each molecule consisting of one old and one new DNA strand (Fig. 5.8).

The properties of the newly synthesised DNA duplex are identical to the DNA which was used as a templet in both base composition and base sequence. Replication of DNA occurs by way of a 'fork' which proceeds along the DNA molecule which, since the two DNA chains are antiparallel, would suggest that the strands are synthesised by two different mechanisms with one chain growing in the $5' \rightarrow 3'$ direction and the other in the $3' \rightarrow 5'$ direction. However, this is not the case. Both daughter strands grow in the $5' \rightarrow 3'$ direction which reinforces the observation that all known DNA polymerases extend polynucleotide chains only at $3'$-hydroxyl termini.

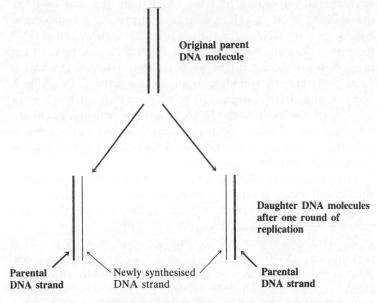

Original parent DNA molecule

Daughter DNA molecules after one round of replication

Parental DNA strand   Newly synthesised DNA strand   **Parental DNA strand**

Fig. 5.8 Schematic representation of semi-conservative replication of DNA

The explanation as to how replication of both strands could be achieved was suggested by Okazaki who proposed that both DNA chains were synthesised by the same mechanism but that one chain was made 'backwards' in short fragments which could be subsequently joined together by DNA ligase which is an enzyme capable of catalysing the formation of a phosphodiester bond between the free 5'-phosphate end of one polynucleotide and the free 3'-OH group of a second polynucleotide positioned next to it. Thus nuclear DNA synthesis proceeds continuously on the 'forward arm' of a replication fork and discontinuously on the 'retrograde arm' (Fig. 5.9)

In eukaryotic cells, DNA polymerase-α is the enzyme solely responsible for nuclear DNA replication and the activity of this enzyme can be consistently correlated with cellular DNA replication. Mitochondrial DNA replication is the responsibility of DNA polymerase-γ but in contrast to nuclear DNA replication, no short DNA intermediates are observed in mitochondrial DNA replication. Mitochondrial DNA is synthesised as a long DNA molecule in a continuous fashion.

In addition to the actual polymerase enzyme, other DNA-binding proteins are involved in the replication process. Single-stranded DNA-binding proteins which may function as helix-destabilising proteins (HD proteins) and prevent reannealing of complementary DNA strands by binding to single-stranded DNA have been isolated from many eukaryotic cells including yeast and fungi. Some of these proteins promote denaturation of DNA while some preferentially stimulate α-polymerase activity. In addition, single-stranded DNA-dependent ATPases which catalyse the unwinding of double-stranded DNA have been detected in eukaryotes, including *Lilium*, the plant enzyme having been shown to promote a limited unwinding of duplex DNA dependent on concomitant hydrolysis of ATP.

The initiation of Okazaki fragment synthesis occurs at random sites within a variable stretch of single-stranded DNA, termed the 'initiation zone' between the replication fork and the 5'-end of a growing daughter strand (Figs. 5.9 and 5.10).

The synthesis begins with the production of an RNA primer of uniform size (8–10 nucleotides) but not of a unique sequence. RNA primer formation is an essential prerequisite for subsequent DNA synthesis since all DNA polymerases require both a templet and a 3'-OH nucleotide primer before they can initiate DNA synthesis. DNA polymerase-α is responsible for synthesising the

135

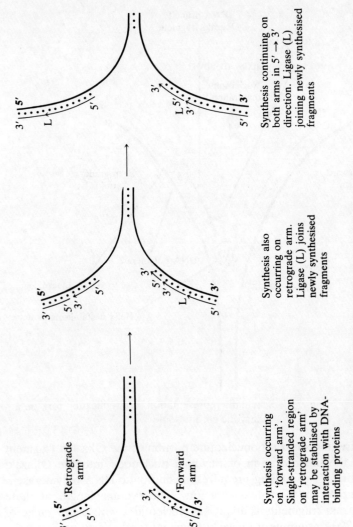

Synthesis occurring on 'forward arm'. Single-stranded region on 'retrograde arm' may be stabilised by interaction with DNA-binding proteins

Synthesis also occurring on retrograde arm. Ligase (L) joins newly synthesised fragments

Synthesis continuing on both arms in 5'→3' direction. Ligase (L) joining newly synthesised fragments

Fig. 5.9 DNA synthesis via Okazaki fragment formation

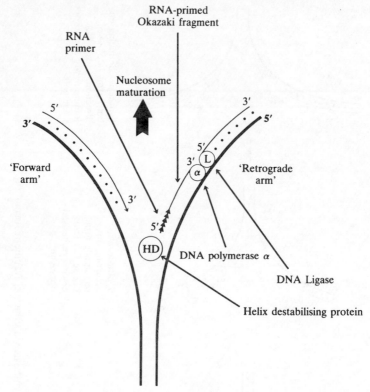

Fig. 5.10 Schematic representation of a replication fork on a
eukaryotic chromosome

entire oligodeoxyribonucleotide portion of the Okazaki fragment
but the incorporation of the final nucleotide into the Okazaki
fragment ('gap-filling') requires in addition to DNA polymerase-α
other essential protein cofactors. The average length of these
Okazaki fragments is about 135 nucleotides with a size range of
40–290 nucleotides.

When the final nucleotide has been inserted, DNA ligase I joins
the two ends of adjacent polynucleotide chains together by formation of a phosphodiester linkage (Fig. 5.10). RNA primers
appear to be removed via a two-step mechanism at about the
same rate that Okazaki fragments are joined to the growing
nascent DNA strands. The majority of the RNA primer is
removed by a ribonuclease-H which endonucleolytically removes
all ribonucleotides except the one linked to DNA, this ribonucleotide being removed by a 5'-3' exonuclease which may be

complexed with DNA polymerase-α and which leaves a 5'-phosphate terminated DNA. It is thought that this second step may sometimes involve the removal of two or more ribonucleotides.

Reassembly of the nucleosome At the same time as replication of the DNA molecule is taking place there is also occurring an unwinding of nucleosomal DNA in front of the replication fork and a reassembly of nucleosomes behind the fork. The first nucleosome on the forward arm is about 125 nucleotides from the 3'-end of nascent DNA while the first nucleosome on the retrograde arm is about 250 nucleotides from the 5'-end of the first Okazaki fragment. Nascent chromatin closest to the actual site of DNA synthesis differs in structure from mature chromatin and goes through a maturation process as the replication fork advances, this process being completed 5–40 kilobases from the actual sites of DNA synthesis. Nucleosome maturation may involve the stepwise addition of histones to nascent DNA and while there are not large pools of histones in somatic cells, histone synthesis appears tightly coupled to DNA replication in a variety of eukaryotic cells.

Ribonucleic acid

The genetic information contained within the genes in the DNA molecule is expressed by determination of the types of proteins synthesised within the cell. However, the DNA molecule itself is not the direct templet for protein synthesis, these templets being ribonucleic acid (RNA) molecules whose base sequence is complementary to the base sequence of the particular gene responsible for its synthesis. The flow of genetic information in normal cells may be depicted in the following manner:

$$\text{Replication} \Big(\text{DNA} \xrightarrow[\text{Transcription}]{} \text{RNA} \xrightarrow[\text{Translation}]{} \text{PROTEIN}$$

The synthesis of RNA using instructions given by the DNA templet is termed transcription, the RNA templets then specifying the biosynthesis of proteins in the process of translation. Not all RNA molecules are information-carrying molecules for protein synthesis, some RNA molecules being integral parts of the general cellular apparatus utilised for the synthesis of proteins. Cells contain three main types of RNA:

1 Messenger RNA (mRNA) – the templet for protein synthesis. There is a mRNA molecule corresponding to each gene or group of genes being expressed, hence mRNA is a very heterogeneous class of RNA molecules.

2 Transfer RNA (tRNA) – these molecules carry amino acids in an 'activated' form to the ribosome for peptide bond formation in a sequence determined by the order of bases in the mRNA templet. There is at least one tRNA molecule specific for each of the 20 amino acids. tRNA is the smallest of the RNA molecules consisting of about 75 nucleotides and having a molecular weight of about 25,000 daltons.

3 Ribosomal RNA (rRNA) – is the major component of ribosomes which are the ribonucleoprotein particles found in the cytoplasm (and also in mitochondria and chloroplasts) and are the sites of protein synthesis in the cell. Cytoplasmic rRNA species are characterised by their sedimentation behaviour, typically 25S, 18S, 5.8S and 5S in plant cells.

Whereas almost all of the DNA is localised in the nucleus of the cell most of the RNA is located in the cytoplasmic ribosomes in association with protein. Ribosomal RNA comprises about 70 per cent of total cellular RNA, tRNA about 25 per cent and mRNA about 2–5 per cent. RNA molecules are initially synthesised in the nucleus of the cell, usually in a precursor form which is larger than the final mature RNA molecule utilised in protein biosynthesis. These precursor molecules are chemically modified, reduced in size and in many cases complexed with proteins before being transported from the nucleus into the cytoplasm. RNA molecules with similar functions but slightly different properties are also synthesised in mitochondria and chloroplasts where they also participate in protein synthesis.

Structure of RNA

RNA consists of a long unbranched polynucleotide chain in which the adjacent sugar residues are joined by $3' \rightarrow 5'$ phosphodiester bonds and each sugar forms a β-linkage to the N-9 of a purine or the N-1 of a pyrimidine base (Fig. 5.11).

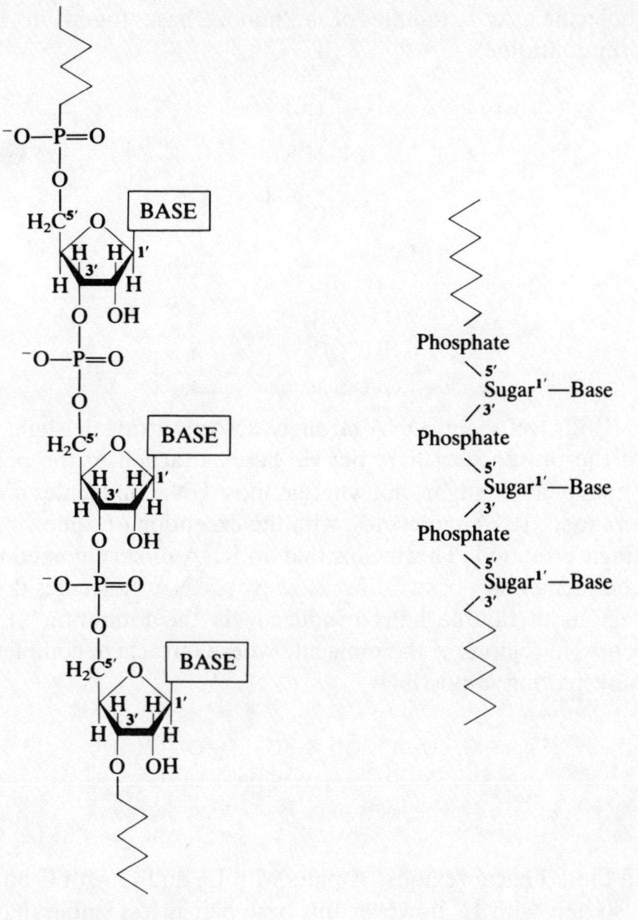

Chemical structure   Simplified general structure

Fig. 5.11 Structure of part of an RNA molecule

The covalent structure of RNA differs from that of DNA in two ways:

1 the sugar unit is ribose, not deoxyribose;
2 one of the four main bases is uracil which replaces thymine. Uracil can form a base pair with adenine.

In addition to the four commonly occurring bases adenine, guanine, cytosine and uracil, other so-called 'rare' or 'minor' bases are found in RNA molecules, particularly tRNA molecules where they can comprise up to 5 per cent of the total base content of the molecule. These rare bases arise from modification of the commonly occurring bases after their polymerisation into the RNA molecule. An example of a 'minor' base found in tRNA is pseudouridine:

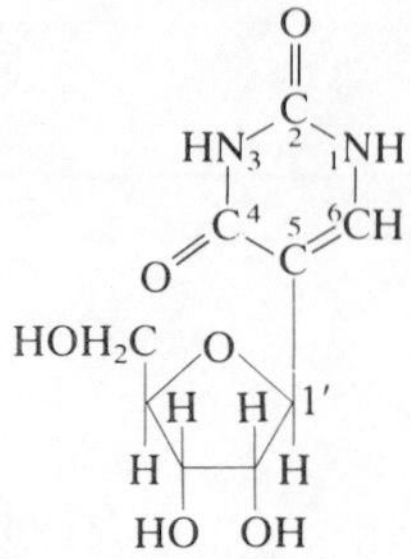

Pseudouridine (5-β-D-ribofuranosyluracil)

Both RNA and DNA strongly absorb ultraviolet light because of the presence of heterocyclic bases attached to the pentose or deoxypentose sugar, but whereas most DNA molecules are double stranded, RNA molecules, with the exception of some viruses, are single stranded. This means that an RNA molecule need not have complementary base ratios. However, RNA molecules do contain regions of double helix produced via the formation of 'hairpin loops' in regions of the molecule where intrachain complementary base pairing is possible:

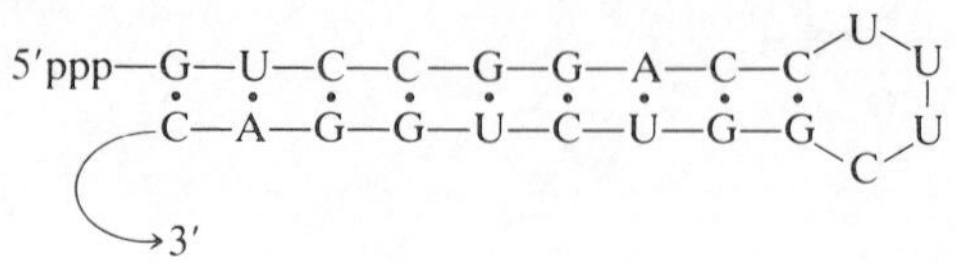

In these helical regions, A pairs with U, and G with C but G may also pair with U, however this base pair is less stable than a G-C base pair. This type of imperfect base pairing is frequently encountered in RNA hairpin loop structures, indeed one or more bases are sometimes looped out to facilitate the pairing of others. The proportion of helical regions found in RNA molecules varies over a wide range.

RNA polymerases

The enzymes responsible for the synthesis of RNA in plant cells are the DNA-dependent RNA polymerases, commonly referred to as RNA polymerases. Plant cells contain three classes of nuclear RNA polymerases plus RNA polymerases of mitochondrial and chloroplast origin, but although there is much information con-

Table 5.5 Properties of nuclear RNA polymerases

Property	Polymerase I	Polymerase II	Polymerase III
Molarity of $(NH_4)_2SO_4$ at which eluted from DEAE – Sephadex column	0.1 M	0.2 M	0.3 M
Effect of α-amanitin	Unaffected	Inhibited by very low concentrations	Inhibited by relatively high concentrations
Intracellular localisation	Nucleolus	Nucleoplasmic	Nucleoplasmic
Transcription products	rRNA precursors	mRNA precursors*	Low molecular weight RNAs*, e.g. tRNA, 5S RNA

* No direct proof exists for plant enzymes but these products are synthesised by the very similar animal polymerases.

cerning properties of the nuclear enzymes there is little information on the properties of the mitochondrial and chloroplast
enzymes. The nuclear RNA polymerases can be distinguished by
their different chromatographic behaviour, sensitivities to the
fungal toxin α-amanitin, their different subunit structures, their
intracellular localisation and by the types of genes which they
transcribe. These properties are summarised in Table 5.5. The
enzyme terminology, i.e. I, II and III, is derived from the order of
elution from DEAE – Sephadex columns using $(NH_4)_2SO_4$ as the
eluting salt gradient.

Plant nuclear RNA polymerases possess two or three high
molecular weight subunits which are different in each of the three
polymerases. The enzymes also differ in the composition of their
small subunits although some of these subunits are common to all
three enzymes (Table 5.6). Wheat germ RNA polymerase II has
been shown to contain approximately seven tightly bound Zn
atoms per enzyme molecule which appear essential for activity of
the enzyme.

Little is known of the mechanisms whereby transcription of the
eukaryotic genome may be controlled but it has been suggested
that modifications to chromatin structure may govern the acces-
sibility of sites on the genome available for transcription by RNA
polymerase. Control may also be exerted at the level of the
enzyme molecule itself and it is interesting to note that alterations
in the large subunit of RNA polymerase II have been observed in

Table 5.6   Subunit structures of RNA polymerases I, II and III from wheat embryo

Subunit molecular weights ($\times 10^{-3}$)

Polymerase I	Polymerase II	Polymerase III	
200	220	150	Large
125	140	130	subunits
		100	
38	42 + 40	55	
		38	
24	27 + 25	28	
	21	25	
20	20	20	Subunits
17.8	17.8	17.8	believed to
17	17	17	be commonly
	16.3		shared
	16		
	14		

tissues undergoing a transition from a quiescent to a metabolically active state. In the case of germinating wheat embryos this alteration in RNA polymerase II involves a specific proteolytic process which converts the 220,000 molecular weight subunit into one with a molecular weight of 180,000. RNA polymerases isolated from quiescent tissues and which possess this relatively larger 220,000 molecular weight subunit are designated RNA polymerase IIA while those from actively metabolising tissues possessing the 180,000 molecular weight subunit are termed RNA polymerase IIB. The appearance of the IIB form in the wheat embryo during germination can be correlated with the activation of RNA synthesis suggesting that the IIA form of the enzyme may be a storage form which is activated for transcription by a specific proteolysis. Additional alterations in the smaller subunit components have also been noted during this conversion of the IIA form into the IIB form during germination.

RNA synthesis

RNA synthesis resembles DNA synthesis in many respects. The RNA chain is synthesised in the $5' \rightarrow 3'$ direction and in the elongation process there is a nucleophilic attack by the 3'-OH group at the terminus of the growing RNA chain on the innermost nucleotidyl phosphate of the incoming nucleoside triphosphate (Fig. 5.12). The RNA polymerase utilises a double stranded DNA templet copying one of the DNA strands, however, in contrast to DNA polymerase the RNA polymerase does not require a primer to initiate polynucleotide synthesis.

In eukaryotes, different polymerases are used to synthesise the different types of RNA (Table 5.5) and initiation of transcription must occur at a specific initiation site on the DNA templet strand in a process which involves local unwinding of the DNA helix and specific interaction between polymerase and an initiation site. The RNA polymerase then copies the DNA strand in the 3' → 5' direction (Fig. 5.12) thus synthesising the RNA chain in the 5' → 3' direction by sequential addition of nucleoside monophosphate units, the order of addition of nucleotides being determined by the base sequence of the DNA templet strand. As elongation of the RNA chain continues the process of chemical modification of the RNA molecule begins (e.g. modification of common bases to produce minor bases), a process which also continues after synthesis is completed. Synthesis stops at a particular site on the DNA molecule suggesting that certain 'stop signals' must be present in the genome and the completed RNA molecule continues to be 'processed' or 'modified' and may be complexed with proteins before being transported from the nucleus into the cytoplasm. The base sequence of the final RNA product is complementary to the base sequence of the gene in the DNA templet copied by the RNA polymerase (Fig. 5.12).

tRNA molecules

The sequence of bases in over 70 tRNA molecules from several different species has been determined and from these studies it has been possible to observe several features common to all these molecules. The tRNA molecule is a single polynucleotide chain containing between 73 and 93 ribonucleotides among which are many of the 'rare' or 'modified' bases examples of which are listed in Table 5.7, typically between 7 and 15 of these modified bases being present in a tRNA molecule.

As can be seen many of these bases are methylated derivatives of the common bases and they are formed by enzymic modification of a precursor tRNA molecule. Such methylation of bases may

144

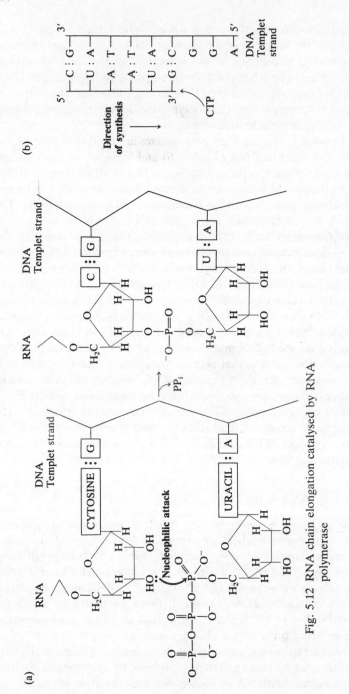

Fig. 5.12 RNA chain elongation catalysed by RNA polymerase

Table 5.7   Examples of minor bases found in RNA

1-Methyladenine	Dihydrouracil
2-Methyladenine	5-Hydroxyuracil
6-Methyladenine	2-Thiouracil
1-Methylguanine	5-Methyl-2-thiouracil
2-Methylguanine	3-Methylcytosine
2,2-Dimethylguanine	5-Hydroxymethylcytosine
Hypoxanthine	2-Thiocytosine
Xanthine	4-Acetylcytosine

restrict the formation of certain base pairs within the molecule thus directing the formation of specific pairing between non-methylated bases while also conferring a hydrophobic character to certain regions of the tRNA molecule which may be important in the interaction of tRNA molecules with synthetase enzymes and ribosomal proteins.

All the tRNA sequences elucidated can be written out in a clover-leaf pattern (Fig. 5.13) in which approximately 50 per cent of the residues are base paired. The 5'-end of the tRNA molecule is phosphorylated and the 5'-terminal residue is usually pG. The base sequence at the 3'-end is usually CCA, these three bases not being involved in hydrogen bonding and the 3'-OH group of the terminal adenosine residue being the site at which the tRNA molecule attaches to the activated amino acid in protein biosynthesis. In addition to the unpaired bases at the 3'-CCA terminus four other groups of bases are unpaired, these being found in the 'looped out' regions of the clover-leaf model – the TψC loop, the DHU loop, the anticodon loop and the 'extra arm' which contains a variable number of residues (Fig. 5.13).

The anticodon loop appears to consist of seven bases having the sequence:

5'—Pyrimidine—Pyrimidine—X—Y—Z—Modified purine—Variable base—3'
$\underbrace{}$
ANTICODON

The clover-leaf model serves to indicate the features common to many tRNA molecules but tells us nothing of the three-dimensional structure of the molecule itself. The polynucleotide chain of the tRNA molecule folds into an 'L'-shape (Fig. 5.14) in which there are two segments of double helix each containing about 10 base pairs corresponding to one turn of the helix. There are other unusual H-bonding interactions between bases in the non-helical regions, e.g. G-G, A-C and these forces along with

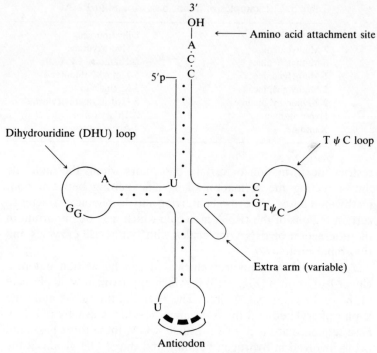

Fig. 5.13 Clover-leaf structure of a tRNA molecule showing common features

other interactions between bases and the 2′-OH of the ribose in the sugar-phosphate backbone and base-stacking forces in helical regions contribute significantly to stabilising the 3-D conformation of the tRNA molecule. The anticodon loop is at the opposite end of the 'L' from the amino acid attachment site (CCA terminus), and since the CCA terminus and adjacent helical region do not interact strongly with the rest of the molecule, it is thought that this part of the molecule may undergo conformational changes during amino acid activation and during protein synthesis when tRNA is bound to the ribosome.

Some tRNA molecules are not synthesised in their final mature state but are synthesised as precursor molecules containing additional nucleotide sequences which must be specifically removed during a post-transcriptional event known as 'processing'. It is during this processing that the 3′-CCA terminus is added to all tRNA molecules. The yeast tRNA[Tyr] precursor has been shown to have extra nucleotide sequences, one at the 5′-end (16 nucleotides), one just following the anticodon (an intervening

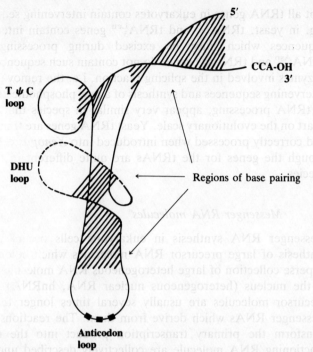

Fig. 5.14 Schematic diagram of 3-D structure of yeast
phenylalanyl-tRNA

sequence of 14 nucleotides) and two additional nucleotides at the
3′-terminus. This precursor is processed into mature tRNA^{Tyr} as
described in the scheme:

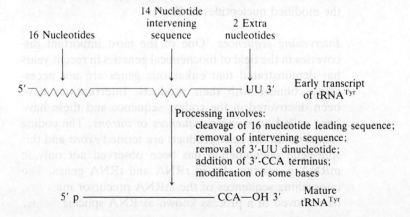

16 Nucleotides

14 Nucleotide
intervening
sequence

2 Extra
nucleotides

5′ ⌇⌇⌇⌇⌇⌇⌇⌇⌇⌇⌇⌇⌇⌇ ⌇⌇⌇⌇ UU 3′　Early transcript
of tRNA^{Tyr}

Processing involves:
cleavage of 16 nucleotide leading sequence;
removal of intervening sequence;
removal of 3′-UU dinucleotide;
addition of 3′-CCA terminus;
modification of some bases

5′ p ——————————————— CCA—OH 3′　Mature
tRNA^{Tyr}

Not all tRNA genes in eukaryotes contain intervening sequences, e.g. in yeast, tRNA^{Tyr} and tRNA$_3^{Leu}$ genes contain intervening sequences which must be excised during processing while tRNA^{Asp} and tRNA$_3^{Arg}$ genes do not contain such sequences. The enzymes involved in the splicing reaction, i.e. the removal of the intervening sequences and synthesis of a new phosphodiester bond in tRNA processing, appear very similar in species that are far apart on the evolutionary scale. Yeast tRNA genes are transcribed and correctly processed when introduced into *Xenopus* cells even though the genes for the tRNAs are quite different in the two species.

Messenger RNA molecules

Messenger RNA synthesis in eukaryotic cells occurs via the synthesis of large precursor RNA molecules which belong to a disperse collection of large heterogeneous RNA molecules found in the nucleus (heterogeneous nuclear RNA, hnRNA). These precursor molecules are usually several times longer than the messenger RNAs which derive from them. The reactions which transform the primary transcription product into the mature functioning RNA molecule are collectively described under the heading RNA processing and these reactions include:

1 exonuclease- and endonuclease-catalysed reactions which alter the size of the transcript;
2 addition of nucleotides either to the 5'-terminus (capping) or to the 3'-terminus (addition of polyadenylic acid or the -CCA addition);
3 enzymes which introduce methyl groups, etc., to produce the modified nucleotides in RNA.

Intervening sequences   One of the most important discoveries in the field of biochemical genetics in recent years has demonstrated that eukaryotic genes are not necessarily colinear with their products. Interruptions have been discovered in the coding sequence and these have been called *intervening sequences* or *introns*. The coding sequences in the gene products are termed *exons* and this split-gene phenomenon has been observed not only in mRNA genes but also in rRNA and tRNA genes. The intervening sequences of the mRNA precursor must first be removed in a process known as RNA splicing before

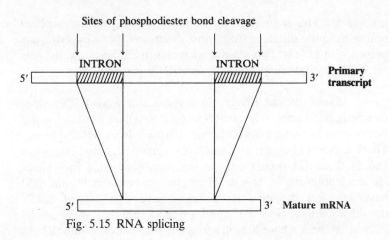

Fig. 5.15 RNA splicing

the mRNA molecule can participate in protein synthesis. This splicing process is performed by RNA splicing enzymes in the nucleus, the intervening sequences being sectioned out and the exposed ends of the exon regions of the mRNA rejoined to form the mature RNA molecule (Fig. 5.15).

It is not known how many RNA splicing enzymes exist in the eukaryotic nucleus but it is possible that both endonuclease and ligase activities reside in the same enzyme. One fact that is known is that splicing enzymes must be very precise in their action. If a one-nucleotide slip in the splice point occurred then this would shift the reading frame of the mRNA molecule on the 3'-side of the splice point and so produce an entirely different amino acid sequence in the protein synthesised which could have serious consequences for the cell in which this occurred. The factors controlling which phosphodiester bonds are broken during splicing are unclear but the base sequences of a number of intron – exon junctions of mRNA transcripts are known and appear to have properties in common. In particular, the base sequence of the intron begins with GU and ends with AG:

Although these characteristics have yet to be shown to be present in plant mRNA precursors, this same splicing signal has been shown to be present in chicken, rabbit and mouse mRNA precursors. The sequences surrounding tRNA precursor splice-points are quite different from those described above implying the presence of at least two splicing enzymes in eukaryotic cells, one for mRNA production and one for tRNA production.

Small nuclear RNAs   Low molecular weight RNA species designated small nuclear RNAs (snRNAs) are localised in the nuclei of eukaryotic cells and could play a role in RNA splicing. There appear to be six major snRNA species designated U1 → U6 and of these U3 occurs only in the nucleolus while the others are nucleoplasmic in origin. They contain between 90 and 220 bases, possess modified nucleotides, have a 5'-cap of 2,2,7-trimethylguanine, (Fig. 5.17) and are most abundant in metaboli-cally active cells where they are found in association with hnRNA. Of particular interest has been the observation that the 5'-end of U1 (the most abundant snRNA) has a sequence exhibiting exten-sive complementarity to the intron–exon boundary sequences of hnRNA and it has been suggested that if the 5'-end of U1 base paired with both ends of the intron (Fig. 5.16) that the remainder of the intron would 'loop out' aligning the two exons for the ligation step. This could be one way in which the specificity of the RNA splicing reaction is achieved.

Modifications at the 5'-terminus of mRNAs (capping)
Eukaryotic mRNAs, including plant mRNAs have a modified 5'-terminus. A methylated guanosine residue is joined to the mRNA by an unusual 5'-5' pyrophosphate linkage (Fig. 5.17) and this highly distinctive structure is introduced into the molecule while the mRNA precursor is being synthesised. The 5'-triphosphate end of the nascent hnRNA chain is hydrolysed to a diphosphate and accepts a guanylate unit from GTP. The guanine base is then methylated by S-adenosylmethionine, which can also methylate the 2'-OH groups of the ribose sugars in animal cells as indicated in Fig. 5.17. However, these ribose units do not appear to be methylated in those plant systems which have been studied. These 'caps' enhance the stability of mRNAs by protecting the 5'-ends from phosphatases and nucleases and also appear to enhance their efficiency of translation in protein synthesis.

Modifications at the 3'-terminus of mRNAs   In addition to chemical modification of internal sugars and bases most cyto-plasmic plant mRNAs are modified at the 3'-terminus by the post-transcriptional addition of a polyadenylic acid (poly A)

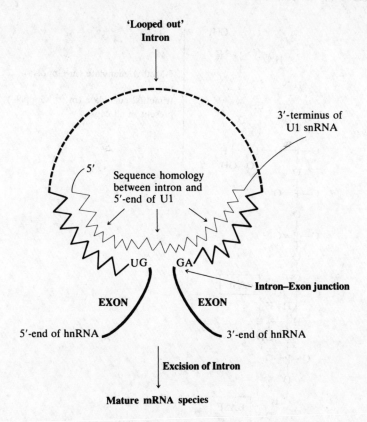

Fig. 5.16 Possible role for snRNA in RNA splicing

'tail' – a region up to 200 nucleotides long. Poly(A) polymerising enzymes and poly(A) hydrolytic enzymes have been isolated from both nuclear and cytoplasmic fractions of plant tissues and although the regulation of mRNA-poly(A) turnover is poorly understood it is possible that the same plant enzyme is capable of poly(A) synthesis and degradation and that its activity in either a synthetic or degradative fashion may be controlled by the cellular ATP level, high levels of ATP favouring synthesis and low levels favouring degradation.

The role of the poly(A) tail in mRNA function is still unclear but it may stabilise the mRNA molecule against nuclease degradation possibly by allowing the mRNA to bind the so-called poly(A)-binding protein which is known to protect mRNA. Not all mRNA species possess a poly(A) tail, one notable exception being histone mRNA which also lacks intervening sequences.

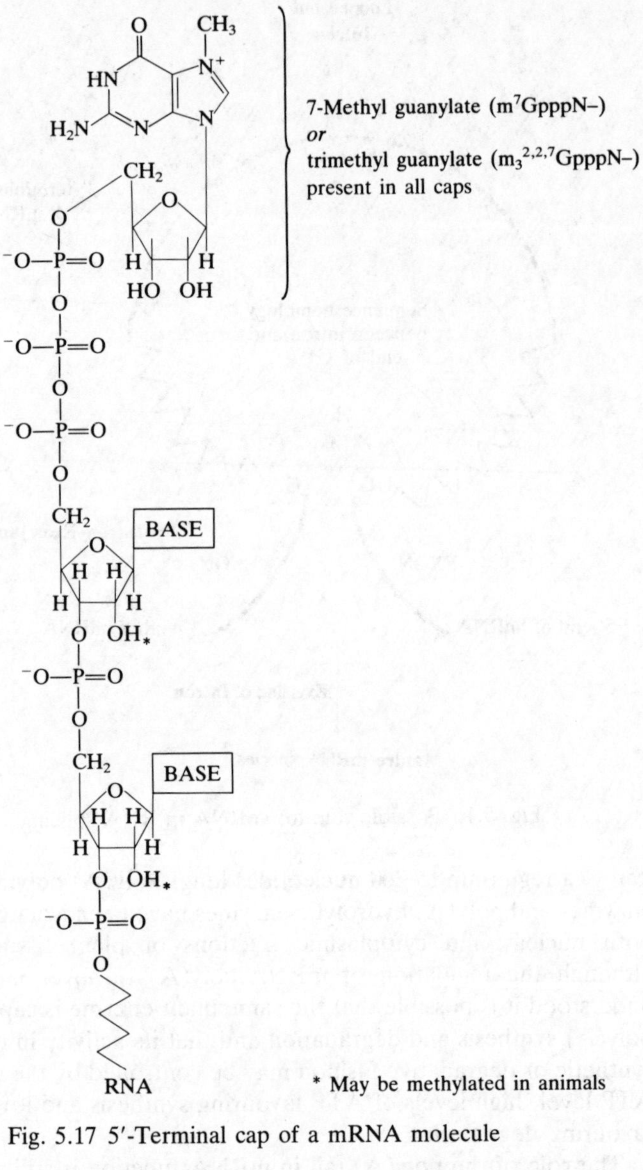

7-Methyl guanylate (m^7GpppN–)
or
trimethyl guanylate ($m_3^{2,2,7}$GpppN–)
present in all caps

RNA   * May be methylated in animals

Fig. 5.17 5'-Terminal cap of a mRNA molecule

mRNA – protein associations   From the very early stages in transcription, hnRNA exists in a complex with protein and free hnRNA has not been detected in cells. Although no clear picture exists concerning the interrelationship of the protein and RNA of this ribonucleoprotein complex, some enzyme activities have been detected in these hnRNP particles including poly(A) synthesising enzymes, 'capping' enzymes and ribonucleases. A group of three or four proteins appear universally in those hnRNPs studied, the proteins being basic proteins which are thought to be the structural proteins of the particle. The modifications such as capping, methylation, splicing, etc., probably occur on this RNP particle, splicing being the terminal step before transport from nucleus to cytoplasm. Little is known about the control of nuclear – cytoplasmic transport except that it involves RNA complexed with protein but this transport of mRNA species from nucleus to cytoplasm could be a point of regulation in the control of gene expression in cells.

Ribosomes and rRNA

The rRNA of the cell is contained in a complex with many proteins in a ribonucleoprotein particle (the ribosome) which is the site of protein synthesis in the cell. In the plant cell, ribosomes can be found in the cytoplasm, either free or membrane bound, in mitochondria and in chloroplasts. These ribosomes from different intracellular locations share many common features although they differ in some of their physical and chemical properties. Ribosomes have a spherical to ellipsoid shape and are composed of two subunits which differ in size, shape, chemical composition and function in protein synthesis. Dissociation of the ribosome into its subunits occurs naturally during the translation cycle and can be induced *in vitro* by altering ionic conditions. The highly ordered structure of the ribosomal subunits is determined basically by specific interactions between the RNA, protein and bivalent cations (mainly Mg^{2+}) which comprise the molecular components of the ribosome.

Cytoplasmic ribosomes   The plant cytoplasmic ribosome is composed of at least 80 different proteins and 4 molecules of RNA (Table 5.8). The large ribosomal subunit has a sedimentation coefficient of 60S and contains 3 molecules of RNA usually referred to by their sedimentation coefficients 26S RNA, 5.8S RNA and 5S RNA together with at least 45 different proteins. The small subunit (40S subunit) contains 1 molecule of 18S RNA and between 35 and 40 proteins. The 26S RNA, 18S RNA and 5.8S RNA are synthesised initially as a large rRNA precursor molecule,

Table 5.8 Properties of plant cell ribosomes

Location	S value monosome	S value subunits	S value	rRNA species Molecular weight (daltons)
Cytoplasm	80S	40S	18S	0.7×10^6
		60S	26S	1.3×10^6
			5.8S	4.5×10^4
			5S	3.9×10^4
Chloroplast	70S	30S	16S	0.55×10^6
		50S	22S	1.1×10^6
			5S	3.9×10^4
			4.5S	$\sim 3.2 \times 10^4$
Mitochondria	78S	40S	18S	0.7×10^6
		60S	25S	1.2×10^6
			5S	3.9×10^4

the first stable rRNA precursor detectable in plant nuclei having a molecular weight in the range $2.2 - 2.6 \times 10^6$ daltons. In some plants more than one rRNA precursor can be detected, e.g. in sycamore cells precursors of molecular weight 3.4×10^6 and 2.5×10^6 have been detected suggesting that the smaller precursor is derived from a larger precursor by a specific processing mechanism.

The DNA containing the genes for cytoplasmic rRNA is associated with the nucleolus where RNA polymerase I, the enzyme responsible for rRNA synthesis, is localised. The DNA which comprises the rRNA genes can be considered to consist of three basic elements:

1 sequences which correspond to those in mature rRNA molecules;

2 sequences which are transcribed as part of an initial precursor molecule but which are not found in mature rRNAs ('transcribed spacer sequences');

3 additional 'non-transcribed spacer sequences' interspersed among the transcribed sequences.

Although the order of the different rRNA molecules in plant rRNA precursors is uncertain, by analogy with animal cell rRNA precursors we might expect the order shown in Fig. 5.18(a) with the 18S rRNA sequence located close to the 5'-end of the precursor and the 26S rRNA towards the 3'-end. The ribosomal RNA genes appear to be tandemly repeated along the ge-

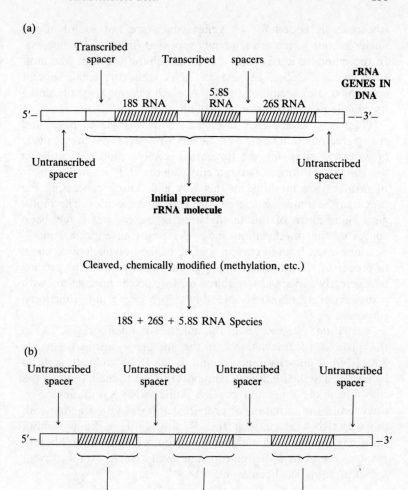

Fig. 5.18 (a) Arrangement of rRNA genes in the DNA genome
(b) Tandemly repeated rRNA genes in the DNA genome

nome separated by untranscribed spacer regions as shown in Fig. 5.18(b). In plant cells approximately 75 per cent of the initial precursor rRNA molecule is conserved in the mature rRNA species whereas only 55 per cent of the precursor is conserved in birds and mammals. The 5S RNA species found in cytoplasmic ribosomes is coded for by genes which are not found in the nucleolus but which are tandemly repeated (in animal genomes). In the monomeric ribosome the large subunit contains one molecule each of 26S, 5.8S and 5S RNA while the small subunit contains one molecule of 18S RNA, each subunit also containing its complement of ribosomal proteins.

Plant cytoplasmic rRNAs contain both methylated bases and sugars, methylation being a post-transcriptional event taking place at selected sites on the precursor rRNA molecule utilising S-adenosylmethionine as the methyl donor. Only a small number of residues are modified in this way and during processing the precursor becomes associated with protein molecules. The biological significance of this methylation is unclear but it has been suggested that methylation and other post-transcriptional modifications assist both in correct folding of the polynucleotide chain of precursor rRNA thus enabling nucleases to process the precursor correctly and are also required for the specific interactions with proteins during ribosome assembly to produce a fully functional ribosome.

Very little is known about the function of eukaryotic rRNAs in the process of translation. In the mature ribosomal particle 5.8S RNA is non-covalently bound to 26S RNA and is thought to be involved in the binding of aminoacyl-tRNA to the 'A' site of the ribosome in the elongation process, while 5S RNA is thought to be involved in the initiation of protein synthesis by interacting with initiator tRNA at or near the 'P' site. Little is known about ribosomal proteins from higher plants but within several different plant species a high degree of evolutionary conservation of ribosomal proteins has been found.

Chloroplast and mitochondrial ribosomes Ribosomes and their subunits from mitochondria and chloroplasts generally have lower sedimentation coefficients, smaller RNA molecules and fewer proteins than cytoplasmic ribosomes (Table 5.8). In addition, although 5S RNA is present in the ribosomes of these plant organelles, no 5.8S RNA is present. However, it has recently been discovered that chloroplast ribosomes do contain an additional small rRNA, 4.5S RNA, associated with the large ribosomal subunit. Methylation of rRNA has also been observed in mitochondrial and chloroplast ribosomes although in mitochondria only the large subunit rRNA appears methylated, there being no methylation in the small subunit. Mitochondrial and chloroplast ribosomes do not share the same ribosomal proteins either with each other or with cytoplasmic ribosomes. Chloroplast and mitochondrial genomes contain the ribosomal genes responsible for the synthesis of organelle rRNAs. An interesting discovery in mitochondria is that the gene for the large rRNA appears to be a 'split' gene containing intervening sequences in the structural gene which are not expressed in the final gene product.

The genetic code

The genetic information stored in the cell as an ordered sequence of bases in the DNA molecule is transferred into a specific sequence of bases contained in the messenger RNA molecule during the process of transcription. Finally, this genetic information is translated from a sequence of bases in mRNA into a specific sequence of amino acids in a protein molecule in the process of protein synthesis. Thus the sequence of bases contained in the gene determines the amino acid sequence of the protein.

The 'code' for each amino acid resides in a sequence of three nucleotides, the codon, found in the mRNA molecule. As there are 64 possible triplet combinations of the four bases adenine, guanine, cytosine and uracil found in mRNA it can be seen that these are more than adequate to code for the incorporation of the 20 naturally occurring amino acids into a protein molecule. This genetic code relating the sequence of bases in the codon to the amino acid incorporated into protein in response to the codon is shown in Table 5.9. Of the 64 possible triplets, 61 have been assigned to particular amino acids while 3 codons, UAA, UAG and UGA which do not code for the incorporation of any amino acids, act as terminating signals causing the synthesis of a polypeptide chain to cease. Many of the 20 amino acids are coded for by more than one triplet and the code is said to be degenerate.

The molecule which acts as the intermediary or adaptor linking the triplet sequence of bases in the codon with the specific amino acid to be polymerised into the growing polypeptide chain in response to that particular codon is the transfer RNA molecule. The tRNA molecule contains an amino acid attachment site to which the amino acid is esterified – the 3'- or 2'-OH of the ribose unit at the 3'(-CCA) end of the tRNA chain – and a templet recognition site consisting of a sequence of three bases known as the anticodon which recognises a complementary sequence of three bases in the codon on the mRNA molecule. Since the code is

Table 5.9 The genetic code

Codon	Amino acid	Codon	Amino acid	Codon	Amino acid	Codon	Amino acid
AUU		ACU		AAU	Asn	AGU	Ser
AUC	Ile	ACC		AAC		AGC	
AUA		ACA	Thr	AAA	Lys	AGA	Arg
AUG	Met	ACG		AAG		AGG	
GUU		GCU		GAU	Asp	GGU	
GUC	Val	GCC	Ala	GAC		GGC	Gly
GUA		GCA		GAA	Glu	GGA	
GUG		GCG		GAG		GGG	
CUU		CCU		CAU	His	CGU	
CUC	Leu	CCC	Pro	CAC		CGC	Arg
CUA		CCA		CAA	Gln	CGA	
CUG		CCG		CAG		CGG	
UUU	Phe	UCU		UAU	Tyr	UGU	Cys
UUC		UCC	Ser	UAC		UGC	
UUA	Leu	UCA		UAA	Ter	UGA	Ter
UUG		UCG		UAG		UGG	Trp

degenerate it follows that more than one kind of tRNA molecule may code for the same amino acid and from Table 5.9 it can be seen that where several codons specify the same amino acid these codons usually differ in only the last (3'-end) base in the triplet.

Until very recently the genetic code was thought to be universal but this no longer seems to be the case as differences have been found in the mitochondrial genetic systems of mammalian cells, yeast, *Neurospora crassa* and higher plants.

Certain variations exist within the codons ascribed to particular amino acids in the cytoplasmic protein-synthesising systems or mitochondrial protein-synthesising systems of these organisms. Human mitochondria read the codon UGA not as a stop codon (see Table 5.9) but as a tryptophan codon while AUA is read not as an isoleucine codon but as a methionine codon. Similarly UGA is read as a tryptophan codon in yeast and *Neurospora crassa* mitochondria. There is tentative evidence for variability in the reading of the genetic code in plant mitochondria but these findings have yet to be confirmed.

Suggestions for further reading

General

Adams, R. L. P., Burdon, R. H., Campbell, A. M., Leader, D. P. & Smellie, R. M. S. (1981) *The Biochemistry of the Nucleic Acids* (9th edn). Chapman & Hall: London & New York.

Bryant, J. A. (1976) *Molecular Aspects of Gene Expression in Plants.* Academic Press.

Burdon, R. H. (1976) *RNA Biosynthesis*, Outline Studies in Biology Series. Chapman & Hall: London.

Hall, T. C. & Davies, J. W. (1979) *Nucleic Acids in Plants*, Vols. I and II. CRC Press Inc.

Smith, H. (1977) *The Molecular Biology of Plant Cells*, Botanical Monographs, Vol. 14. Blackwell Scientific Publications.

Specific

DNA

Bedbroke, J. R. & Kolodner, R. (1979) Chloroplast DNA, *Ann. Rev. Plant Physiol.*, **30**, 593–620.

Depamphilis, M. L. & Wassarman, P. M. (1980) Replication of eukaryotic chromosomes, *Ann. Rev. Biochem.*, **49**, 627–66.

McGhee, J. D. & Felsenfeld, G. (1980) Nucleosome structure, *Ann. Rev. Biochem.*, **49**, 1115–56.

RNA

Abelson, J. (1979) RNA processing and the intervening sequence problem, *Ann. Rev. Biochem.*, **48**, 1035–69.

Bielka, H. & Stahl, J. (1978) Structure and function of eukaryotic ribosomes, in H. R. V. Arstein (ed.) *International Review of Biochemistry; Amino Acid and Biosynthesis II*, Vol. 18. University Park Press: Baltimore.

6

Protein biosynthesis

In the process of translation the genetic information contained in the linear sequences of ribonucleotides of the messenger RNA molecules is converted into the linear sequence of amino acids which make up the different enzymic and structural proteins of the cell. Plant cells contain three different protein synthesising systems which are to be found localised in the cell cytoplasm, mitochondria and chloroplasts. The mitochondrial and chloroplast protein-synthesising systems are very similar but they differ significantly from the cytoplasmic system in that the components of the system in these organelles bear a greater similarity, from a functional viewpoint, to the components of prokaryotic systems rather than to those of the eukaryotic cytoplasm. For the purpose of discussing the reactions involved in protein biosynthesis, the process can be conveniently divided into several distinct steps:

1 amino acid activation and aminoacyl-tRNA synthesis;
2 peptide chain initiation;
3 peptide chain elongation;
4 chain termination, peptide and ribosome release from the polysome complex.

Each of these steps will now be considered in turn. The reactions of the eukaryotic cytoplasmic system will be discussed initially and this will be followed by comparison with the mitochondrial and chloroplast systems.

Amino acid activation, aminoacyl-tRNA synthesis

The first step in protein biosynthesis involves the activation of amino acids in the presence of ATP, Mg^{2+} ions and enzymes known as aminoacyl-tRNA synthetases (amino acid:tRNA ligases) to produce an enzyme-bound aminoacyl-adenylate or activated amino acid. The same synthetase enzyme then catalyses the transfer of the amino acid from the aminoacyl-adenylate to a specific tRNA molecule. In the aminoacyl-tRNA molecule the amino acid is esterified to the tRNA via the 2′- or 3′-hydroxyl group of the terminal (3′-) adenosine residue. There appears to be only one aminoacyl-tRNA synthetase enzyme for each of the naturally occurring amino acids while in some instances more than one tRNA species for a given amino acid is known. The overall reaction can be represented as a two-step process as illustrated in Fig. 6.1.

The process of protein biosynthesis is characterised by the absolute specificity with which the correct amino acids are incorporated into each position in the polypeptide chain. The aminoacyl-tRNA synthetase must discriminate between potential amino acid substrates and between the many tRNA species in the cell, and it can be envisaged that severe problems might arise in distinguishing between structurally similar amino acids such as alanine and glycine, isoleucine and valine. It has been suggested that an 'editing process' exists whereby the aminoacyl-tRNA synthetase selects the correct amino acid for activation and transfer to tRNA via sites on the enzyme which select for both correct molecular size and chemical characteristics and hence allow the enzyme to discriminate between closely related amino acids.

There are no reports of naturally coexisting amino acids being attached to the wrong tRNA by aminoacyl-tRNA synthetases under physiological conditions, an observation which confirms that such an 'editing process' must exist. Indeed, certain plants contain naturally occurring amino acid analogues and in these plants the synthetase enzyme discriminates against these amino acids in the amino acid activation reaction, whereas the corresponding synthetase from plants in which the amino acid analogue is absent does not exhibit this discrimination mechanism, and can activate the analogue and in some cases esterify the analogue to tRNA. Examples of naturally occurring amino acid analogues which are discriminated against by the synthetase of their host plants are azetidine carboxylic acid, a proline analogue, which is found in high levels in *Polygonatum multiflorum* (Solomon's seal), canavanine, an arginine analogue occurring naturally in *Canavalia ensiformis* (Jack bean) and mimosine, a naturally occurring phenylalanine analogue found in *Mimosa* and *Leucaena* (Fig. 6.2).

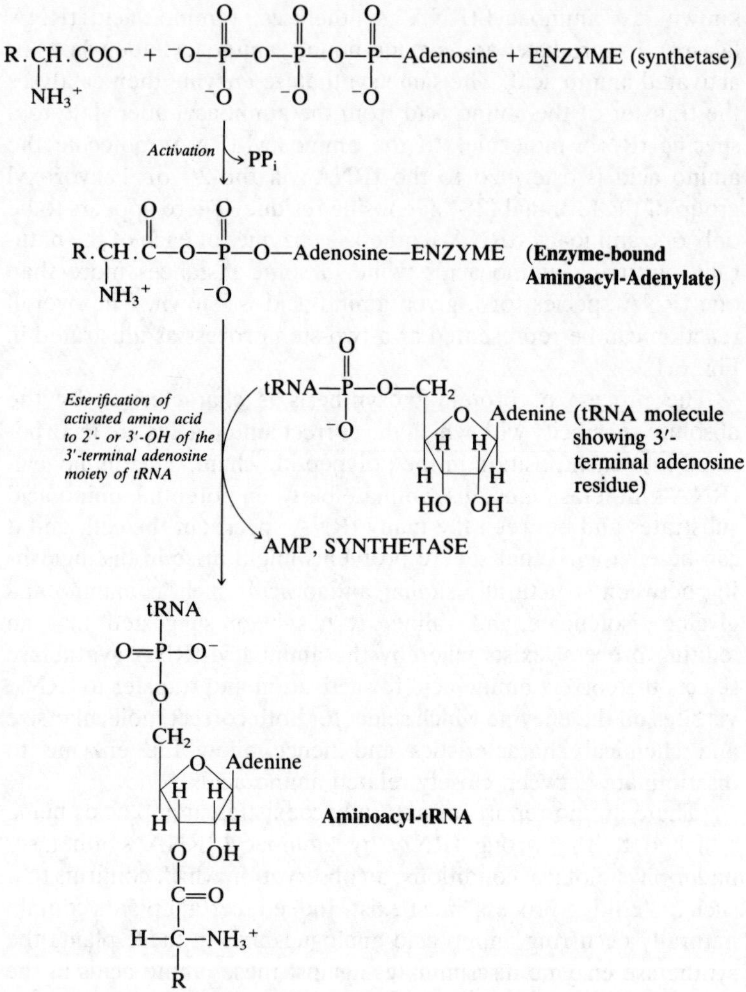

$$R.CH.COO^- + {}^-O\!-\!\overset{\displaystyle O}{\underset{\displaystyle O^-}{P}}\!-\!O\!-\!\overset{\displaystyle O}{\underset{\displaystyle O^-}{P}}\!-\!O\!-\!\overset{\displaystyle O}{\underset{\displaystyle O^-}{P}}\!-\!Adenosine + ENZYME\ (synthetase)$$

with NH$_3^+$ on the amino acid.

Activation $\searrow$ PP$_i$

$$R.CH.\overset{\displaystyle O}{C}\!-\!O\!-\!\overset{\displaystyle O}{\underset{\displaystyle O^-}{P}}\!-\!O\!-\!Adenosine\!-\!ENZYME$$

with NH$_3^+$ on the amino acid.

(Enzyme-bound Aminoacyl-Adenylate)

Esterification of activated amino acid to 2'- or 3'-OH of the 3'-terminal adenosine moiety of tRNA

tRNA—P(=O)(O$^-$)—O—CH$_2$ ... O Adenine (tRNA molecule showing 3'-terminal adenosine residue) with H H, H H, HO OH

$\downarrow$ AMP, SYNTHETASE

tRNA
O=P—O$^-$
O
CH$_2$
O Adenine
H H
H H
O OH
C=O
H—C—NH$_3^+$
R

Aminoacyl-tRNA

Fig. 6.1 Amino acid activation

Once the tRNA molecule has been charged with the correct amino acid it is now equipped to perform its function as an adaptor molecule to locate the amino acids in the correct positions for peptide synthesis as dictated by the mRNA in the polysome complex. In addition, after peptide bond formation the tRNA molecule binds the growing polypeptide chain to the ribosomes.

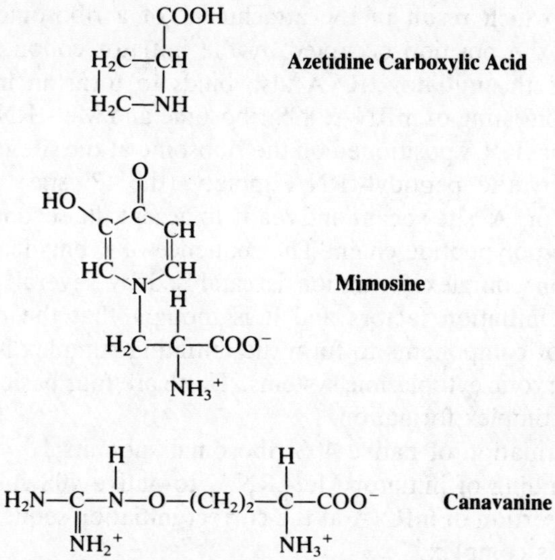

Fig. 6.2 Examples of amino acid analogues which occur in plants

Peptide chain initiation

Natural messenger RNA molecules contain a specific initiation codon (AUG) which is recognised by a specific initiator tRNA species. Initiation in all organisms involves a methionine-tRNA species and in the cell all proteins contain an N-terminal methionine residue (N-formyl methionine in prokaryotic systems) although in many cases the methionine residue is lost during post-translational modification of the protein. There are at least two species of tRNA which can be charged with methionine present in the cells of all organisms. The initiator-tRNA, designated $tRNA_f^{Met}$ (Met-tRNA$_f$ when charged with the amino acid), is involved only in the initiation steps of protein biosynthesis whereas the other methionine-accepting tRNA species, $tRNA_m^{Met}$, is involved only in the elongation steps and donates methionine residues 'internally' into the growing polypeptide chain. Both $tRNA^{Met}$ species read the AUG codon in mRNA but $tRNA_f^{Met}$ also reads GUG and UUG, an example of 5'-'wobble' in codon recognition which is very unusual as it is usually the third, i.e.3'-base of the codon which is concerned in this phenomenon.

   The initiation steps of protein biosynthesis involve a series of interactions between various components of the protein synthetic apparatus which result in the attachment of a ribosome to the mRNA at the position occupied by the initiator codon. During these steps the initiator tRNA also binds to form an initiation complex consisting of mRNA, 80S ribosome and Met-tRNA$_f$ with the initiator tRNA positioned on the ribosome at the site normally occupied by the peptidyl-tRNA moiety (the 'P' site) and the aminoacyl or 'A' site vacant and ready to accept the second amino acid in the polypeptide chain. The sequence of events leading up to initiation complex formation is catalysed by several proteins known as initiation factors and it is thought that the order of additions of components to form the initiation complex is similar in all eukaryotic cytoplasmic systems. There are four basic steps in initiation complex formation:

    1 formation of native 40S ribosomal subunits;

    2 binding of initiator Met-tRNA$_f$ to native 40S subunits;

    3 insertion of mRNA at the correct initiation sequence into this complex;

    4 joining of the 60S ribosomal subunit to form a completed 80S initiation complex

The individual steps of this pathway are illustrated schematically in Fig. 6.3. Although the functional characterisation of all the protein factors required for 80S initiation complex formation in higher plants is still lacking some comparison is possible with other eukaryotic systems which have been studied in more detail.

Formation of native ribosomal subunits

Ribosomes can only bind to mRNA as separate subunits and the 80S ribosomes formed during chain termination transiently dissociate into free subunits upon their release from mRNA but under *in vivo* conditions non-enzymic reassociation into nonfunctional 80S monomers is favoured. However, in the cell these subunits are prevented from joining together by a protein which acts as an anti-association factor and in eukaryotes is termed initiation factor 3 (eIF-3). This factor binds to the smaller ribosomal subunit independently of an energy supply or other factors and prevents the larger subunit from binding (Fig. 6.3, step (i)). A factor with the properties of eIF-3 is present in wheat germ where the large eIF-3 complex contains 9 − 11 subunits ranging in molecular weight from 35,000 to 160,000. In addition wheat germ also possesses an anti-association factor eIF-6 which appears to bind to the 60S ribosomal subunit and prevents its reassociation

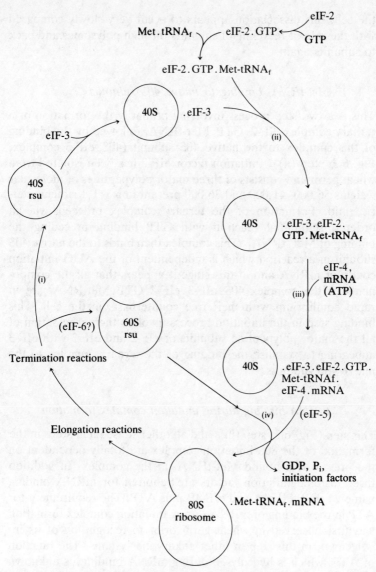

Fig. 6.3 Initiation steps in protein biosynthesis

with the small subunit. Both these factors eIF-3 and eIF-6 appear to prevent reassociation of subunits rather than actively dissociating 80S ribosomal couples, but most eukaryotic cells still contain a sizeable pool of 80S ribosomes which are incapable of participating in protein synthesis unless they can be dissociated into subunits. In the cell, this dissociation appears to occur very slowly compared with the rate of recycling of subunits through polysomes and back to subunits again.

Met-tRNA$_f$ binding to native 40S subunits

This is a two-step process involving primarily the formation of a ternary complex eIF-2.GTP.Met-tRNA$_f$ followed by the binding of this complex to the native 40S subunit (40S.eIF-3 complex, Fig. 6.3, step (ii)). Initiation factor eIF-2 has been isolated from wheat germ and consists of three major polypeptides of molecular weights 56,000, 41,000 and 36,000 present in a 1:1:1 stoichiometric ratio. Formation of the ternary complex proceeds via an ordered sequential reaction with GTP binding preceding the binding of Met.tRNA$_f$. This complex then binds to the native 40S subunit in a reaction which is independent of the AUG initiation codon or mRNA and at this stage it appears that all the components of the complex 40S.eIF-3.eIF-2.GTP.Met-RNA$_f$ are in rapid equilibrium with their free counterparts in the cell. This binding step in the initiation process involves the participation of all the major polypeptide subunits of eIF-2 and eIF-3 with eIF-3 appearing to stabilise the binding of the ternary complex to the native 40S subunits.

mRNA binding during initiation complex formation

This step (Fig. 6.3, step (iii)) and all other subsequent steps in the formation of the 80S initiation complex are totally dependent on the presence of bound Met-tRNA$_f$ in the complex. In addition three distinct initiation factors are required for mRNA binding namely eIF-3, eIF-4A and eIF-4B plus ATP. The requirement for ATP in the binding of mRNA during initiation complex formation was first observed in wheat germ prior to recognition of its involvement in this step in other eukaryotic systems. The function of ATP, which is hydrolysed during mRNA binding, is unknown although it might be involved in an energy-dependent translocation which would align the anticodon of the 40S-bound Met-tRNA$_f$ with the initiator codon AUG of the mRNA. Just as wheat germ has been shown to possess initiation factors eIF-2 and eIF-3 it has also been shown to possess eIF-4B, a polypeptide of molecular weight 80,000. At the end of this stage of initiation complex formation there exists the pre-initiation complex 40S ribosomal subunit . eIF-3 . eIF-2 . GTP . Met-tRNA$_f$. eIF-4 . mRNA (Fig. 6.3) awaiting binding of the 60S ribosomal subunit. The exact details of the mechanism by which eukaryotic ribosomes bind at the initiator AUG codon are not fully understood, but it is possible that the ribosomes attach to a position on the messenger to the 5'-side of the coding sequence prior to alignment of the AUG codon with the initiator-tRNA.

Joining of the 60S ribosomal subunit This step requires initiation factor eIF-5 and involves a concerted set of reactions which result in Met-tRNA$_f$ becoming bound at the 'P' site of the 60S subunit, GTP being hydrolysed to GDP and P$_i$ and all the initiation factors leaving the complex (Fig. 6.3, step (iv)). The primary function of the bound GTP appears to be one which allows an energy-dependent eIF-5 mediated release of eIF-2 and eIF-3 prior to subunit joining. Following release of these factors the remaining complex is left in a state such that if 60S ribosomal subunits are available then formation of a functional 80S initiation complex will occur whereas if joining of the 60S subunit is precluded then the destabilised 40S ribosomal subunit. Met. tRNA$_f$. mRNA complex probably dissociates.

Peptide chain elongation

The protein chain is synthesised from its N-terminus by the sequential addition of amino acids in the process of peptide chain elongation. This repetitive process begins with the growing polypeptide chain esterified to tRNA (i.e. as peptidyl-tRNA) occupying the donor or 'P' site on the ribosome and the acceptor or 'A' site unoccupied (Fig. 6.4). The aminoacyl-tRNA species whose anticodon is complementary to the mRNA codon adjacent to the peptidyl-tRNA now binds to the 'A' site (Fig. 6.4, step (i)). Peptidyl transferase, a component of the large ribosomal subunit then catalyses the formation of a peptide bond between the esterified carboxyl group of the amino acid of peptidyl-tRNA and the α-amino group of the amino acid of the aminoacyl-tRNA at the 'A' site resulting in the production of an elongated (by one amino acid residue) peptidyl-tRNA species now bound at the 'A' site (Fig. 6.4, step (ii)). The uncharged tRNA which had initially bound the peptidyl moiety then dissociates from the 'P' site and the elongated peptidyl-tRNA is translocated from the 'A' site to

168

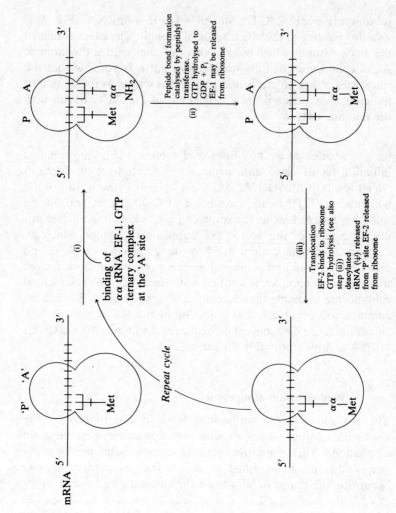

Fig. 6.4 Elongation steps in protein biosynthesis

the now vacant 'P' site as the ribosome and mRNA molecule move by the length of one codon relative to each other (Fig. 6.4, step (iii)). The situation is now analogous to that at the beginning of the elongation cycle except that a new codon on the mRNA has been exposed at the 'A' site hence a new aminoacyl-tRNA can bind at this position. The elongation steps are repeated many times in the above manner during the synthesis of a protein.

The overall result of one round of the elongation process is the extension of the nascent polypeptide chain by one amino acid residue. In eukaryotic cells this process requires the presence of two elongation factors, termed EF-1 and EF-2. Peptide bond formation between peptidyl-tRNA and aminoacyl-tRNA might be expected to occur spontaneously since both are reactive species, however, peptide bond formation is catalysed by an enzyme peptidyl transferase which is a protein component of the large ribosomal subunit. The excess of energy released during peptide bond formation and in the hydrolysis of GTP occurring during the elongation process appears not to be required for actual bond formation but it might be that this energy is utilised during the translocation step of the elongation cycle in which the ribosome and mRNA move relative to each other in an energy-requiring process.

Binding of aminoacyl-tRNA to the ribosome

The binding of non-initiator aminoacyl-tRNA to the 'A' site on the 80S ribosome is mediated via elongation factor 1 (EF-1), a factor which has been isolated from a variety of eukaryotic cells. In mammalian systems the active monomeric form of EF-1 is designated $EF-1_L$ or $EF-1\alpha$ and consists of a single polypeptide chain which tends to aggregate with itself and with other polypeptide complexes which might mediate the recycling of EF-1 during its role in aminoacyl-tRNA binding to the ribosome. There is some evidence that $EF-1\alpha$ is associated with a complementary factor $EF-1\beta$ which appears to mediate the release of $EF-1\alpha$. GDP from the ribosome after aminoacyl-tRNA binding and facilitates exchange of GTP for the GDP in this binary complex.

In plant systems in which the elongation process has been studied it has been shown that EF-1 can exist in many different forms which have different molecular sizes. Heavy species of EF-1 having molecular weights in the range 240,000–540,000 daltons cannot spontaneously change into lighter forms neither can they form ternary complexes with GTP and aminoacyl-tRNA. However, a ternary complex between a lighter form of EF-1 (EF-1$_L$, molecular weight 67,000 daltons), GTP and aminoacyl-tRNA has been detected and results suggest that binding of aminoacyl-tRNA to the 'A' site of 80S ribosomes in plant tissue follows the scheme shown in Fig. 6.5. The heavy form of EF-1 first forms a complex with GTP prior to conversion into a lighter form, subsequently this binary EF-1$_L$. GTP complex binds aminoacyl-tRNA which subsequently becomes bound at the 80S ribosomal 'A' site. Upon GTP hydrolysis an EF-1. GDP complex must be released from the ribosome and recycled as postulated in Fig. 6.5, with GTP being exchanged for GDP if EF-1 is to participate in another round of the elongation cycle. The relative affinities of GDP and GTP for EF-1 in plant systems are not known but in mammalian systems EF-1α appears to have a greater affinity for GTP than for GDP, the exchange of GTP for GDP in the binary complex being facilitated by EF-1β. No such analogous factor as EF-1β has been demonstrated in plant systems.

Translocation

The translocation of peptidyl-tRNA from the ribosomal 'A' site to the 'P' site with a simultaneous movement of the ribosome a distance of one codon towards the 3'-end of mRNA is mediated by EF-2 and accompanied by release of deacylated tRNA from the 'P' site and GTP hydrolysis (Fig. 6.5). This hydrolysis of GTP is not required for binding of EF-2 to the ribosome but is a prerequisite for release of EF-2 after translocation. Wheat germ EF-2 is a single polypeptide chain having a molecular weight of about 70,000 daltons, the stoichiometric ratio of EF-2: ribosomes being approximately 1.0 in this tissue. Whereas EF-1 from wheat germ contains no cysteine residues, EF-2 does contain some cysteine and sulphydryl groups appear essential for translocase activity. However, the exact mechanism by which the process of ribosome translocation along mRNA occurs is not known. Similarly the number of GTP molecules hydrolysed per peptide bond formed remains a point of uncertainty and although each of the elongation factors seems to require GTP hydrolysis for activity *in vitro* the possibility remains that *in vivo* only one GTP molecule may be hydrolysed per elongation cycle. Although EF-1 and EF-2 activities appear to exist separately in unmoistened wheat germ (10 per cent moisture content) they form an EF-1. EF-2 complex on

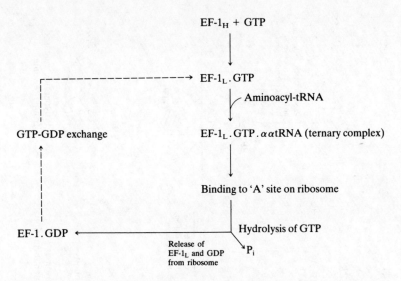

(---→ Postulated steps)

Fig. 6.5 Aminoacyl-tRNA binding to 80S ribosomes

moistening. It has been suggested that *in vivo* these elongation factors may be associated in a complex of this nature and that the formation of the complex may be dependent upon the physiological state of the plant material, in particular the hydration conditions.

Termination of peptide chain synthesis

Chain termination in eukaryotic systems appears to be basically similar to that in bacteria, however, evidence concerning the detailed mechanism of the process in plant systems is lacking. The termination process in mammalian system is outlined in Fig. 6.6 and can be considered to involve two distinct processes:

    1 terminator codon recognition;
    2 the hydrolysis of peptidyl tRNA on the 'P' site of the ribosome.

A similar mechanism of termination may also be operative in plant systems.

Terminator codon recognition

In the termination process the codon recognition molecules are proteins not tRNA (contrast initiation and elongation see p. 163).

172

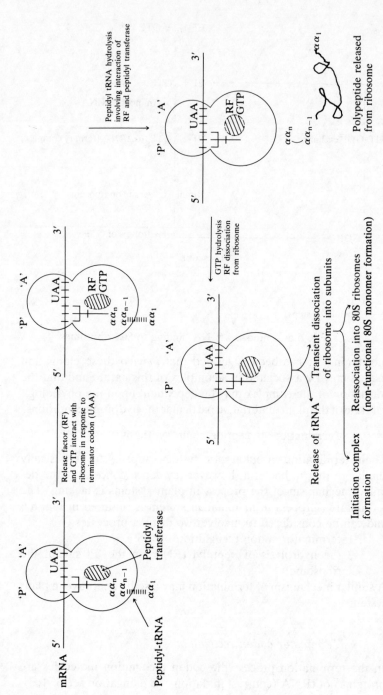

Fig. 6.6 Termination steps in protein biosynthesis

Recognition of the terminator codons UAA, UAG and UGA in the mRNA molecule requires a protein release factor designated RF which will function with any of the three terminator codons and which recognises the terminator codon directly. It has been suggested that GTP facilitates binding of RF to the ribosomes possibly in the interface between the two ribosomal subunits and RF may interact not only with the terminator codons but also with the 3'-end of 18S rRNA.

Hydrolysis of peptidyl-tRNA

After the binding of the carboxy-terminal aminoacyl-tRNA to the ribosome and subsequent translocation from the 'A' site to the 'P' site the terminator codon is now aligned at the 'A' site. Since there is no aminoacyl-tRNA having a complementary recognition anti-codon to the terminator codon, RF and GTP bind to the terminator codon 'A' site region as described in the previous section. It is proposed that RF and peptidyl transferase then interact and the peptidyl transferase which normally catalyses peptide bond formation now hydrolyses the ester bond linking the peptide chain to the tRNA molecule in peptidyl-tRNA resulting in the release from the ribosome of a completed protein chain. Simultaneous hydrolysis of GTP leads to dissociation of the release factor RF from the ribosome. Ribosomes then leave the messenger RNA molecule, possibly as subunits which are then free to participate in another round of protein synthesis or reassociate to form 80S ribosomes which join the cellular 80S 'ribosome pool'.

Post-translational modification of proteins

The polypeptide released from the ribosome at termination may not necessarily be in its final form and may undergo alterations in size via proteolytic cleavage reactions to produce the final protein product. Indeed many proteolytic cleavages may occur while the nascent polypeptide chain is undergoing synthesis and hence still attached to the ribosome, e.g. the initiating methionine residue of eukaryotic proteins is often removed when only 15–30 amino acid residues have been polymerised. Many enzymes may be synthesised and stored in a precursor form which must then be proteolytically cleaved prior to their activation and subsequent involvement in cell metabolism. The secondary and tertiary structure of the polypeptide probably changes continuously throughout synthesis until the final product possesses a structure of minimum energy with most polar groups exposed to the aqueous environment and most non-polar groups in the interior of the molecule. Other alterations to composition which may occur post-translationally are phosphorylation, glycosylation and methylation of amino acid side chains.

Translational control of protein synthesis

There is very little direct evidence concerning control of protein biosynthesis at the level of translation in eukaryotic systems. Any controls which do operate may involve control of the types of proteins synthesised via selection of particular mRNA species for translation (quality control) or control of the frequency of translation of mRNA molecules (quantity control). Alterations in the rate of synthesis or nature of proteins synthesised seem mainly to operate at the initiation step although it is possible that the proteins associated with mRNAs may specifically repress or enhance their translation. Eukaryotic cells also contain a sizeable pool of 80S ribosomes which do not participate in protein synthesis unless they undergo dissociation into subunits. Under certain conditions, e.g. stress conditions, many ribosomes appear to enter this inactive ribosome pool but can equally rapidly leave it during active growth phases. These two situations contrast with the normal very slow dissociation of inactive 'pool' ribosomes into active subunits and this flux of ribosomes between active and inactive states may be an area of translational control although at the present time it is not clear exactly which factors regulate these fluxes.

It is possible that the energy charge or energy state of the cell may regulate translation by regulation of the rate of initiation. Ternary complex ($eIF-2 . GTP . Met-tRNA_f$) formation appears to be a primary site at which the rate of protein synthesis could be adjusted to the energy state of the cell. Binding of $Met-tRNA_f$ to native 40S ribosomal subunits is inhibited by GDP which has a hundred-fold higher affinity for eIF-2 than has GTP. The guanine nucleotide pool in the cell is linked to the larger adenine nucleotide pool by nucleoside diphosphate kinase, thus formation of this ternary complex and hence rate of initiation of protein synthesis can be rapidly co-ordinated with the overall adenylate energy charge of the cell.

Another possible site for translational control of protein synthesis could be the elongation step involving EF-1. Aggregated (heavy, $EF-1_H$) forms of EF-1 which may be held together by phospholipids and cholesterol esters have been shown to exist in eukaryotic cells, including plant cells. It is possible that these aggregated forms of EF-1 may be a storage (inactive) form for the active $EF-1\alpha$ subunit and that the amount of active form available for participation in protein synthesis could be controlled by the association of the active form of EF-1 into, or dissociation of active EF-1 from the aggregated inactive $EF-1_H$ form.

Protein biosynthesis in chloroplasts and mitochondria

Initiation

Mitochondrial and chloroplast protein-synthesising systems exhibit many similarities to prokaryotic systems and are significantly different from the cytoplasmic system. The initiation steps involve a formylated initiator tRNA, N-formyl methionyl-$tRNA_f$, and although characterisation of chloroplast and mitochondrial initiation factors is lacking it might be expected that they will resemble those of bacteria where three initiation factors are involved in initiation steps similar to those found in the eukaryotic cytoplasm. Initiation factor-3 (IF-3) is an 'anti-association' factor which binds to the smaller ribosomal subunit and prevents the larger subunit from binding to it. IF-3 may also be involved in mRNA binding to ribosomes. IF-2 is defined as that factor required for the binding of initiator tRNA to the ribosomes while IF-1 has ill-defined functions but may have activity in the dissociation of ribosomes into subunits and in stabilising the initiation complex.

Elongation

The mitochondria and chloroplasts also contain elongation factors which are different to those of the cytoplasm. These have been studied in more detail than any of the other factors involved in organelle protein synthesis and again have been shown to possess properties resembling those of prokaryotic elongation factors rather than eukaryotic cytoplasmic elongation factors. The two chloroplast elongation factors $EF-T_{Chl}$ and $EF-G_{Chl}$ appear to be synthesised in the chloroplast itself utilising genetic information encoded in the chloroplast DNA. However, in contrast, mitochondrial elongation factors have been shown to be synthesised in the cell cytoplasm and not in the organelles, the genetic information for their synthesis apparently residing in nuclear DNA. The elongation steps in prokaryotic, organelle and eukaryotic cytoplasm appear very similar. Elongation factor-2 (EF-2) of the cytoplasm has properties similar to elongation factor G of bacteria, however, EF-1 of the cytoplasm and EF-T of bacteria do differ somewhat.

If EF-T$_{Chl}$ resembles EF-T of bacteria then it might be involved in a sequence of reactions similar to those shown in Fig. 6.7 for the bacterial system. EF-T consists of two components EF-T$_u$ and EF-T$_s$ present in equimolar amounts. Only EF-T$_u$ is concerned with the codon-directed binding of aminoacyl-tRNA to the vacant ribosomal 'A' site next to the occupied 'P' site while EF-T$_s$ is required only to recycle EF-T$_u$ in an active form. The formation of a binary complex of EF-T$_u$ and GTP precedes binding of aminoacyl-tRNA in the ternary complex (Fig. 6.7) and it is in the form of this complex that the aminoacyl-tRNA binds to the

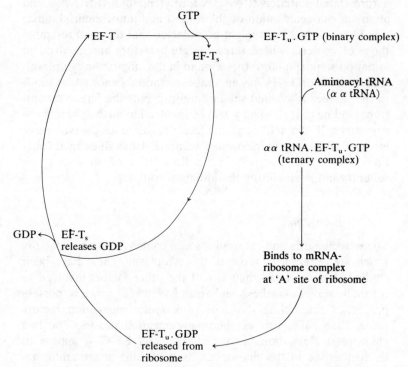

Fig. 6.7 Role of elongation factor T in protein biosynthesis

ribosomal 'A' site. After binding, $EF-T_u$.GDP leaves the ribosome and $EF-T_s$ displaces GDP to regenerate active EF-T.

Termination

It is anticipated that the process of termination of protein synthesis in chloroplasts and mitochondria is similar to the prokaryotic system and is an active process involving protein factors and utilising terminator codons. Whether all three terminator codons UAA, UAG and UGA of bacterial and eukaryotic cytoplasmic systems are used in mitochondria and chloroplast systems in plants remains to be established but certainly other eukaryotic mitochondrial systems can read UGA as a tryptophan codon instead of a terminator codon.

Although the chloroplast synthesises some of its own proteins, e.g. the large subunit of Fraction 1 protein (ribulose bisphosphate carboxylase, which catalyses the initial steps in both photosynthesis and photorespiration) and elongation factors T and G, other chloroplast proteins are synthesised in the cell cytoplasm and must of necessity be transported across the chloroplast membrane prior to assembly into functioning units in the chloroplast itself, possibly via interactions with chloroplast-synthesised proteins. Some products of cytoplasmic protein synthesis which eventually become chloroplast proteins are synthesised as high molecular weight precursor proteins, e.g. the small subunit of Fraction 1 protein. These precursors can be transported across the chloroplast membrane and cleaved to their final mature size within the chloroplast itself. In addition to the several chloroplast-synthesised proteins which have been identified, numerous minor products of chloroplast protein synthesis remain unidentified.

Cytoplasmic synthesis and subsequent transport

Several proteins which are synthesised in the cytoplasm on 80S ribosomes do not have a functional role to play in the cytoplasm itself but are either secreted from the cell, thus having an extracellular role, or are specifically acquired by intracellular organelles, e.g. chloroplasts and mitochondria, where they have a role to play in the structure or function of these organelles. This poses the question as to how the cell recognises which proteins should be secreted by the cell or acquired by intracellular organelles from among the multitude of different protein species synthesised on cytoplasmic ribosomes. Two hypotheses have been forwarded to explain this phenomenon and both postulated mechanisms may be operative in plant cells.

The signal hypothesis

This hypothesis was designed to explain the synthesis and secretion of secretory proteins by cells but is also applicable to the synthesis of storage proteins by developing seeds (see Chapter 7). Both secretory proteins and the storage proteins of seeds appear to be synthesised on ribosomes bound to the endoplasmic reticulum (ER) in cells. The proteins synthesised on these membrane-bound ribosomes are transported across the ER membrane into the lumen of the ER while still being synthesised and then proceed to either storage vesicles or to the extracellular environment of the cell in the case of secreted proteins.

It is proposed that all mRNAs responsible for the synthesis of secretory proteins contain a sequence of bases which codes for the synthesis of unique signal peptide sequence, rich in hydrophobic amino acids, at the N-terminus of the polypeptide. This sequence is present only in nascent incomplete chains and is removed at some stage prior to completion of polypeptide synthesis. The steps involved in the synthesis of a protein for secretion or storage by the cell are illustrated in Fig. 6.8 and are summarised below:

    (i) Translation of the mRNA begins on free cytoplasmic ribosomes.

    (ii) Elongation of the polypeptide chain continues until 10–40 amino acid residues have been added (the signal sequence in the case of secreted or storage proteins).

  (iii) The membrane of the ER now distinguishes between polypeptide chains with or without the unique signal sequence.

  (iv) The mRNA – ribosome complex synthesising those chains possessing the signal sequence binds to the membrane through association with several ribosome receptor proteins present on the ER which form a 'tunnel' in the membrane.

   (v) This association permits passage of the signal peptide through the tunnel into the lumen of the ER.

  (vi) Polypeptide chain synthesis continues as the newly synthesised peptide enters the lumen of the ER at the same time as the signal peptide is hydrolytically cleaved from

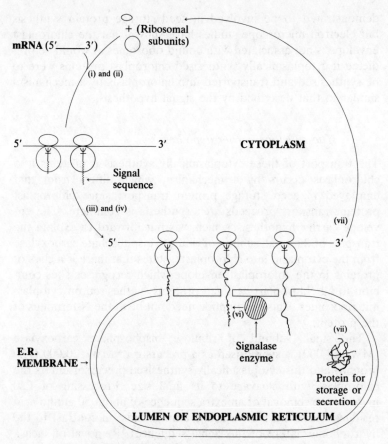

Fig. 6.8 The signal hypothesis (*Note*: (i), (ii), etc., refer to reactions mentioned in text)

the nascent growing polypeptide chain by a protease ('signalase') associated with the ER membrane.

(vii) On completion of protein synthesis the ribosome dissociates from the polysome – membrane complex, the membrane tunnel is eliminated and the ribosome is free to be involved in the translation of any cytoplasmic mRNA species.

As can be seen, an important feature of this hypothesis is that membrane transport of the protein depends on simultaneous protein synthesis by membrane-bound ribosomes thus causing the peptide chain to migrate through the tunnel in the membrane as it is being extended. Membrane-bound ribosomes have been demonstrated to be involved in seed storage protein synthesis but electron microscope studies have shown that the chloroplast envelope is not associated with bound ribosomes as would be predicted if cytoplasmically synthesised chloroplast proteins were to be synthesised and transported into chloroplasts by a mechanism similar to that described by the signal hypothesis.

The envelope carrier hypothesis

The transport of these cytoplasmically synthesised proteins into chloroplasts occurs by a mechanism which differs from that employed in seed storage protein transport since chloroplast protein transport proceeds *after* synthesis is complete. The envelope carrier hypothesis, which was put forward to explain the transport of the small subunit of ribulose bisphosphate carboxylase from the cytoplasm into chloroplasts, states that there is a class of proteins in the chloroplast envelope which recognises sites common to all chloroplast-destined proteins synthesised on cytoplasmic ribosomes. This polypeptide need not be at the N-terminus of the protein.

The small subunit of ribulose bisphosphate carboxylase (MW~14,000) is synthesised as a precursor of MW~20,000 (P20 protein) and this cytoplasmically synthesised precursor enters the chloroplast with cleavage to its final size. Processing of P20 involves the removal of an extra sequence of about 50 amino acid residues which are rich in acidic amino acids (in contrast to the hydrophobic nature of the signal sequence). Removal of such a large number of charged residues causes a conformational change in the protein molecule which leads to transport of the protein across the chloroplast envelope and release of the protein into the stroma.

Thus the envelope carrier hypothesis proposes that:
  (i) the small subunit of ribulose bisphosphate carboxylase is synthesised as a precursor (P20) on cytoplasmic ribosomes;
  (ii) P20 combines with a specific carrier in the chloroplast envelope;
  (iii) envelope-associated protease activity removes a part of the polypeptide which is particularly rich in acidic amino acid residues;
  (iv) this removal of such a large number of charged amino acid residues triggers off a conformational change in the

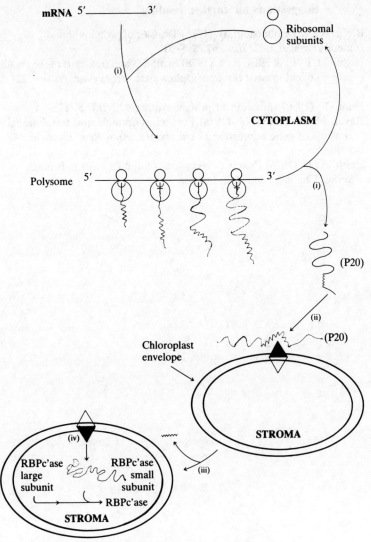

Fig. 6.9 The envelope carrier hypothesis (*Note*: (i), (ii), etc.,
refer to reactions mentioned in text)

polypeptide leading to its transport across the chloroplast envelope and into the stroma.

The steps in this hypothesis are illustrated in Fig. 6.9 and the hypothesis may be applicable to the acquirement of many cytoplasmically synthesised proteins by both mitochondria and chloroplasts.

Suggestions for further reading

Blobel, G. & Dobberstein, B. (1975) Transfer of proteins across membranes, *J. Cell. Biol,* **67**, 835–51.

Highfield, P. E. & Ellis, R. J. (1978) Synthesis and transport of the small subunit of chloroplast ribulose bisphosphate carboxylase, *Nature*, **271**, 420–4.

Hunt, T. (1980) Initiation of protein synthesis, *TIBS*, **5**, 178–81.

Revel, M. & Groner, Y. (1978) Post-transcriptional and translational controls of gene expression in eukaryotes, *Ann. Rev. Biochem.*, **47**, 1079–26.

Smith, A. E. (1976) *Protein Biosynthesis*, Outline Studies in Biology Series. Chapman and Hall: London.

7

Nitrogen interconversions and transport during plant development

Seed germination

Provided that any dormancy has been broken, the trigger which stimulates the germination process seems to be the imbibition of water at an appropriate temperature by cells of the partially dehydrated seed. This stimulus requires the embryo of the mature seed to respond in a co-ordinated manner and within minutes of rehydration taking place many of the metabolic processes which has been arrested or, at best, performed at a greatly reduced rate in the quiescent seed have been reactivated.

The synthesis of RNA and protein commences within minutes of the rehydration of the seed while appreciable DNA synthesis may only be detectable after an initial lag period of several hours during which time several of the enzymes required for DNA replication are activated or synthesised. Some of the genes which were expressed during the period of seed maturation are now repressed during the germination phase while other genes coding for proteins required specifically for the germination process become derepressed. Thus the germination period involves a distinct alteration in genome expression by the seed.

During the early stages of germination many enzyme systems become functional and storage materials within the seed are degraded and subsequently transported from the cotyledons or endosperm to the developing embryonic axis. Seed germination is characterised by a decline in storage tissue nitrogen content and a corresponding increase in the nitrogen content of the embryonic axis during growth with the major decline in endosperm or cotyledon nitrogen content being due to a decline in the protein content of these reserve tissues. The amino acids produced as a result of reserve protein hydrolysis can be transported to the growing embryonic axis where they can be used as a nitrogen source to support growth until the developing seedling is able to make maximal use of the nitrate or ammonia available to it in the soil solution. This changeover from a reliance on reserve tissue-supplied nitrogen to that of an exogenous inorganic source during embryo growth requires the induction of nitrate reductase activity in embryo tissues and relies on a supply of reductant which would be forthcoming when photosynthesis commences during illumination of leaves and cotyledons.

The metabolism of seed storage proteins

The storage proteins of seeds are found primarily in spherical or oval-shaped subcellular organelles known as protein bodies bounded by a single membrane. In addition to storage protein, the other major component of the protein bodies is phytin (the calcium or mixed $Ca^{2+} - Mg^{2+}$ salts of inositol hexaphosphate). The largest amounts of storage reserves are to be found in the endosperm of monocotyledons or the cotyledons of dicotyledons and in most seeds except those of Gramineae the principal storage proteins are globulins. With the exception of rice, the seed protein of commonly grown cereals consists of 40–60 per cent prolamines which contain proline and glutamine as the principal amino acids but are low in lysine, and 20–40 per cent glutelins although these cereals also contain albumins and globulins.

Hydrolysis of the reserve proteins by proteolytic enzymes during germination produces amino acids which serve as a source of carbon and nitrogen for the synthesis of cellular proteins, amino acids and other nitrogen-containing compounds in the developing seedling. Reserve proteins are frequently enriched in specific amino acids, e.g. arginine, proline, glutamine, thus the composition of the amino acid pool resulting from reserve protein hydrolysis does not necessarily reflect the amino acid requirements of the growing seedling. However, the seedling possesses the enzymic machinery to enable it to perform extensive amino acid interconversions, e.g. via transamination and deamination reactions (see pp. 60–64), and so produce an amino pool suitable for its requirements at this time.

Legume storage proteins are glycoproteins and their amino acid hydrolysis products appear to undergo extensive interconversions prior to transport from cotyledons to the developing seedling. In contrast, cereal storage proteins are hydrolysed to amino acids which are probably transported to the growing seedling without extensive modifications. Thus relatively high concentrations of glutamine and proline, which are abundant amino acids in cereal storage proteins, are transported to the seedling where they are metabolised to other amino acids as required. Where extensive amino acid interconversions in reserve tissues do occur prior to transport then the amino acid nitrogen is often channelled into the amido nitrogen of glutamine which then becomes the principal nitrogen-containing compound exported to the developing axis from cotyledon or endosperm tissue. The hydrolysis of reserve proteins and subsequent transport of hydrolysis products from reserve tissue to developing seedling is summarised in Fig. 7.1.

Arginine may serve as a storage reserve in some woody perennials while canavanine, a non-protein amino acid, accumulates in some legume seeds. These amino acids can both be degraded initially via the action of the enzyme arginase to produce either ornithine and urea from arginine, or canaline and urea from canavanine. Ornithine may be transported directly to the developing seedling or further metabolised to glutamate, γ-amino butyrate or proline prior to transport. The enzyme urease converts the urea produced in these reactions to ammonia and carbon dioxide, the ammonia then being utilised in glutamate or glutamine formation and it is in this form in which the urea nitrogen is transported to the developing axis tissue. One of the key enzymes involved in this interconversion, glutamine synthetase, has been shown to accumulate in cotyledons during germination at a time when these pathways would be operative.

Nucleic acid metabolism during seed germination

The nucleic acid content of the reserve tissues of seeds varies widely depending on the type of seed. The endosperm tissue of cereal seed has a relatively low RNA content while the cotyledons of leguminous seed contain much greater quantities of RNA. During germination, the RNA content of the cotyledons of peas and beans declines as the nucleic acid is hydrolysed to nucleotides prior to transport to the developing axis tissue. Similarly the already low RNA content of cereal endosperm declines even further during the germination of cereal seed. The DNA content of cotyledons of peas and beans, however, actually exhibits a slight increase during the early stages of germination as mitochondria

186

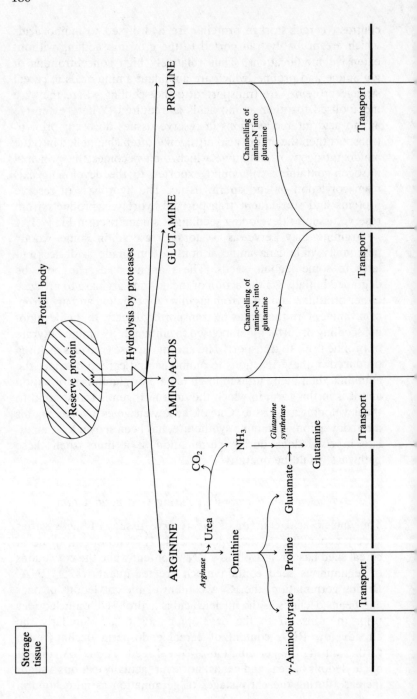

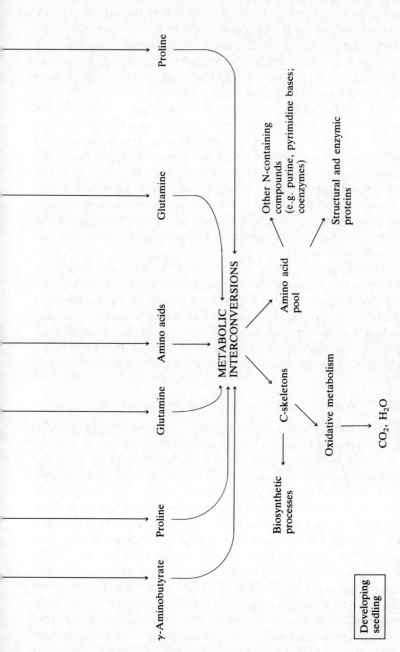

Fig. 7.1 Hydrolysis of reserve proteins during germination

proliferate in the cells of the reserve tissue, this observed increase being in mitochondrial DNA content rather than nuclear DNA content and occurs in the absence of any cell division within the reserve tissue. Seeds which exhibit epigeal germination show an increase in cotyledonary RNA levels when the cotyledons undergo greening on exposure to light, this increase occurring in both chloroplast and cytoplasmic RNA components.

The observed decline in the RNA content of cotyledonary storage tissue during germination does not mean that the machinery for gene expression is totally repressed in these tissues. Indeed RNA synthesis is active during the period in which there is a quantitative decrease in RNA content of the storage organs and the proteins synthesised during this period of germination may contribute both to general cell maintenance and more specifically to the degradation and mobilisation of storage materials in the reserve tissue, e.g. via enzymes such as lipases, proteases, nucleases, amylases, etc.

During the early phase of germination there is some correlation between the disappearance of nitrogenous storage material from reserve tissues and the appearance of the nitrogen-containing degradation products (amino acids, nucleotides) in the axis, although as germination progresses there is an increasing dependence of the developing seedling on *de novo* synthesis of these nitrogenous compounds in axis tissue for use in growth processes.

Nucleotide metabolism during seed germination

In germinating wheat embryos there is a rapid increase in ATP levels from ~20 pmol ATP per embryo in the quiescent state to ~250 pmol ATP per embryo over the first hour of germination. This dramatic increase cannot be wholly accounted for by conversion of the ADP and AMP present in the embryos prior to imbibition into ATP and so indicates a contribution from *de novo* synthesis of the nucleotide. Similar rapid increases in ATP levels on imbibition have been observed in soybean embryonic axis tissue. The levels of other ribonucleoside triphosphates, GTP, CTP and UTP all exhibit similar sharp increases during the first hour of germination of wheat embryos but in the subsequent four hour period of germination it is only the pyrimidine ribonucleoside triphosphate levels which continue to increase significantly (the levels increase four-fold above the one hour level and approach the levels of GTP in the tissue). The purine ribonucleotide levels show only a 60 per cent increase over this period. As the germinating wheat embryo contains significant levels of RNA polymerase activity and since the chromatin of these embryos shows no detectable change in templet activity over this initial germination period, it has been suggested that, while it is the dramatic increase in the levels of both purine and pyrimidine nucleotides which triggers off the initial synthesis of RNA in wheat embryos during the first hour of germination, from one to five hours of germination it is the level of pyrimidine nucleotides alone which controls the rate of RNA synthesis.

In the axis tissue of germinating soybeans there is evidence that the *de novo* synthesis of purine nucleotides can arise from either of two independent pools of IMP which appear to be present in the axis tissue. One of these pools is derived by *de novo* synthesis while the other arises from a salvage pathway and there appears to be little interaction between these pools. Soybean axes also contain nucleotide sugar derivatives during early germination. The dominant class of nucleotide sugars in axis tissue during this period is the UDP-sugars but ADP-sugars and GDP-sugars have also been isolated. However, whereas the levels of ADP-sugars remains unchanged during the early germination stages, GDP-sugars which are virtually undetectable in quiescent axis tissue do show an increase up to the time of commencement of growth.

Nitrogenous transport components

Vascular plants transport nitrogenous solutes from the regions of assimilation in the roots into their aerial shoot portions principally through the xylem while the phloem is used to transport photosynthate from its site of production in photosynthetic tissues to its site of utilisation in other parts of the plant. Nitrogenous compounds of xylem sap are usually present at levels in the range 0.01–0.21 per cent (w/v) and make up the main class of xylem sap components on a dry matter basis. They are also found in relatively high concentrations in phloem sap but on a quantitative basis are second in importance to the large amounts of carbohydrate transported in phloem sap mainly in the form of sucrose.

Plants transport the majority of their nitrogen by using a limited number of general nitrogen transport compounds. A general characteristic of the molecules involved in nitrogen transport is that they should have a low C:N ratio. The use of such compounds ensures the most economic use of the plant's carbon resources in nitrogen transport and also permits the continuing efficient transport of nitrogen under conditions in which the carbon supply is limited.

The restricted range of protein amino acids used in nitrogen transport have two characteristics in common – they all have a low C:N ratio and their carbon skeletons are readily derived from or give rise to oxo acids which are intermediates of the tricarboxylic acid cycle. Of the amide amino acids used, asparagine has C:N ratio of 2.0 and glutamine a ratio of 2.5 while arginine, another amino acid utilised in nitrogen transport, has a C:N ratio of 1.5. In addition to protein amino acids, non-protein amino acids such as γ-methylene glutamine are used by some plants as nitrogen transport compounds while others utilise ureides such as allantoin for this purpose (Fig. 7.2). Ureides have a distinct advantage as nitrogen transport molecules in terms of carbon economy as the C:N ratio in the molecule is 1.0.

The major solutes which a plant uses to transport nitrogen are characteristic of a particular species and in the majority of plants studied the same nitrogen transport compounds are utilised in both xylem and phloem transport systems. The chief nitrogen transport compound in legumes is asparagine but in nodulated plants of *Glycine max*, *Phaseolus vulgaris* and *Vigna unguiculata*, ureides contribute a significant proportion of the fixed nitrogen leaving the nodule via the xylem. In soybean, the nodules may export only ureides thus any amide or amino nitrogen which is transported from the roots to the shoots via the xylem must arise through the assimilation of combined nitrogen by the roots. In contrast to the situation in soybeans, the Vicieae and Trifolieae utilise amides and amino acids as the principal nodule nitrogenous export compounds. Asparagine is also used as a short distance transport compound in trees which additionally utilise arginine both for long distance transport and as a storage form for overwintering. Other amino acids which are utilised in significant amounts as nitrogen transport compounds are glutamine, glutamic acid and aspartic acid.

In certain species, a significant proportion of the xylem-transported nitrogen can be found as alkaloids while other species use non-protein amino acids such as γ-methylene glutamine as in the case of peanuts. The xylem sap of certain species contains nitrate (but not nitrite) and nitrate levels become elevated particularly when high levels of nitrate are available to the root system in

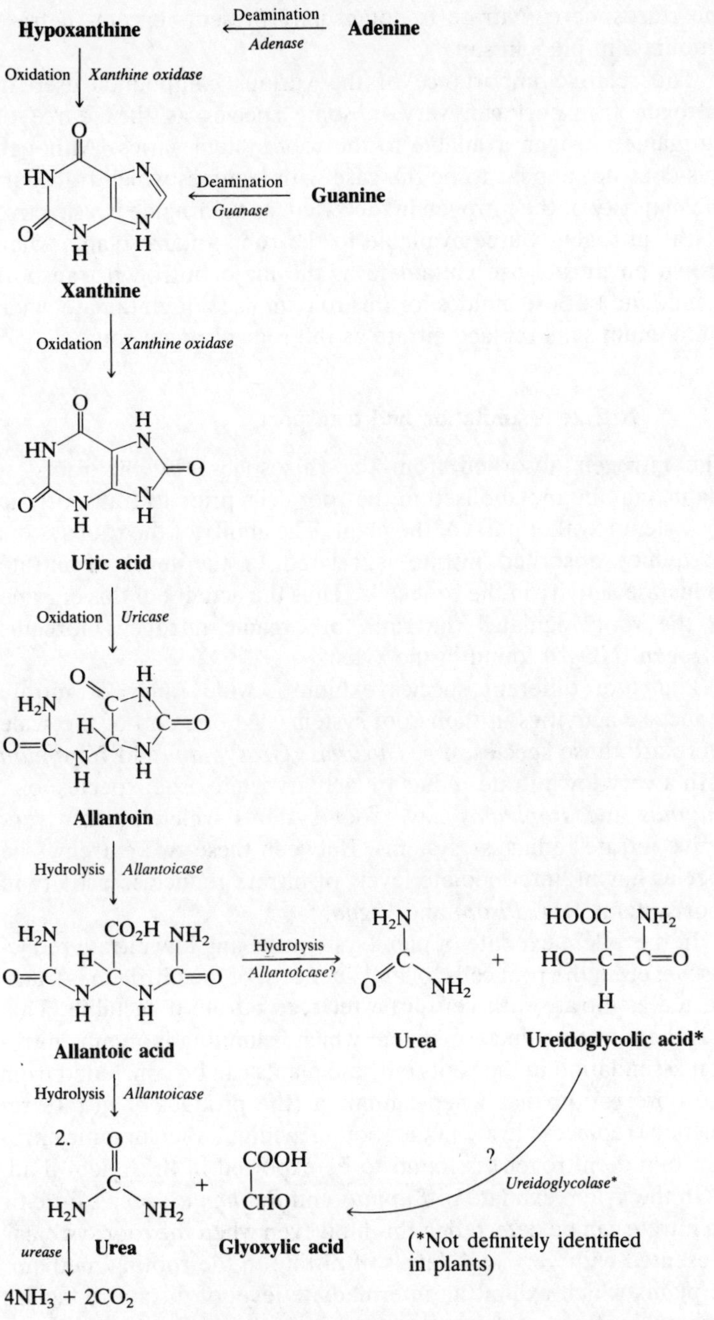

Fig. 7.2 Ureide biosynthesis

the rhizosphere. Nitrate is not usually present, except in trace amounts, in phloem sap.

The relative importance of the various compounds used in nitrogen transport can vary in some species as the source of inorganic nitrogen available to the root system varies. Although this does not appear to be the case with legumes which transport the majority of the nitrogen in the xylem as asparagine irrespective of the nitrogen source available to the roots, maize plants when grown on nitrate use glutamate as the major nitrogen transport compound but use amides for this role in place of glutamate when ammonium salts replace nitrate as the root nitrogen source.

Nitrate assimilation and transport

The nitrogen absorbed from the rhizosphere by the roots of plants may be metabolised in the root cells prior to transport via the xylem to other parts of the plant. The ability of the root system to reduce absorbed nitrate is related to the level of nitrate reductase activity in the root cells. Thus the activity of this enzyme in the root regulates the ratio of organic nitrogen:inorganic nitrogen (NO_3^-) found in the xylem.

Plants of different species exhibit a wide range of nitrate reductase activities in their root systems. At one end of the scale there are those species, e.g. *Cucumis*, *Gossypium* and *Xanthium* with a very low nitrate reductase activity while other species, e.g. *Lupinus* and *Raphanus* have root systems which possess very active nitrate reductase systems. Between these two extremes lie species having intermediate levels of nitrate reductase activity in root systems, e.g. *Pisum* and *Vigna*.

In the xylem exudate of plant systems having low nitrate reductase levels in the root cells, over 95 per cent of the nitrogen is transported as nitrate with very little reduced nitrate detectable. That it is the nitrate reductase system which is limiting inorganic nitrogen assimilation in the roots of these plants can be concluded from the observation that when ammonia (the product of nitrate reduction) replaces nitrate in the rooting medium, then organic nitrogen (amide nitrogen) is found to be exported in the xylem fluid.

In the xylem exudate of *Lupinus* and *Raphanus* species little or no nitrate can be detected in this fluid even when the root system is presented with very high levels of nitrate in the rooting medium. In plants which exhibit an intermediate level of nitrate reductase activity in their roots both free nitrate and organic nitrogen-containing compounds can be found in xylem fluid. The balance between these two forms of nitrogen in the xylem fluid of these plants can be altered by altering the concentration of nitrate available to the plant in the rooting medium. In *Pisum* and *Vigna* the level of organic nitrogen in xylem fluid increases as the nitrate concentration in the rooting medium increases until a maximum level of xylem organic nitrogen is reached at a time at which the root's nitrate reductase system has become saturated. Subsequently, there is a gradual increase in the fraction of xylem nitrogen transported as nitrate. Nitrate reductase activity in roots may also be controlled via the availability of a carbohydrate supply from the shoot system, this supply being used to generate the essential reductant for the nitrate reductase system via oxidation reactions in the respiratory pathways of the root cell.

Much less information is available on the uptake and assimilation of nitrogen by woody species. The form in which nitrogen is transported in the xylem of these species is almost entirely organic, taking the form of asparagine and arginine in *Pyrus*. This indicates that the root system of these woody species possesses the capacity to reduce the inorganic nitrogen (NO_3^-) taken up by the roots from the rhizosphere.

A small proportion of the products of nitrate assimilation in the roots of a plant are not transported out of the roots by the xylem but are used either as a store of nitrogen for future root growth and other metabolic purposes or are transported to growing regions of the root to support ongoing cell division and metabolism within these regions.

The composition of the nitrogenous solutes in the xylem sap changes depending upon both time of day or night and upon the particular period in the life cycle of the plant. The circadian rhythm noted in rates of xylem nitrogen output from roots exhibits a maximum near noon and a minimum at midnight. During daylight, when transpiration rates are high, the transpirational delivery of nitrogen from the roots to the shoots of the plant in the xylem occurs at a faster rate than does translocation of the photosynthate, etc., from the leaves to the roots via the phloem. At night the reverse occurs and a relatively low transpiration rate allows a net translocation of materials from the leaves to the roots during the hours of darkness during which time there is a build up of the 'nitrogen pool' in the roots which in turn is rapidly depleted through xylem transport to the shoot as transpirational activity increases during the following day.

During plant development, the different parts of the plant undergo both quantitative and qualitative changes in their nitrogen requirements. In the initial stages of seedling development the levels of nitrogenous solutes in the xylem are low but there is a rapid increase in these levels when the root system has developed sufficiently to commence metabolism and export of absorbed nitrogen. These high levels of xylem nitrogen are maintained until after the flowering stage when a distinct fall in the amount of xylem nitrogen exported from the root system is observed. In the final stages of the plant's growth cycle, the xylem fluid can contain a wide range of nitrogenous compounds many of which are derived not merely from the metabolism of absorbed nitrogen by the roots but also from the mobilisation of the nitrogen reserves laid down in root tissues. These nitrogenous compounds may be utilised in the formation of seed reserves which are being laid down at this time in the life cycle of the plant.

Nitrogenous transport compounds exported by root nodules

Additional nitrogenous compounds may be found in xylem sap as a result of nitrogen fixation occurring in root nodules. The nodules of plants which are actively fixing nitrogen secrete nitrogenous compounds into the xylem, the types of compound secreted being characteristic of a particular species and are often different from the nitrogenous compounds secreted into the xylem fluid as a result of the assimilation of combined nitrogen (NH_4^+ or NO_3^-) by the root system.

In soybean, it is possible that the nodules export only ureides (allantoin and allantoic acid) and thus any nitrogen transported by the xylem stream from the root system in the form of amide- or amino-N has been produced not by the nodules but through the assimilation of combined nitrogen by the roots. In contrast the nodules of the Trifolieae and Vicieae export amides and amino acids. These differences between the xylem export products of nodules and root tissue become particularly evident when the nitrogenous components of the xylem sap of nodulated plants grown under conditions where the plant is actively fixing N_2 are compared with those of a non-nodulated plant grown on fixed nitrogen (NO_3^-, NH_4^+).

Pisum arvense (field pea) is an amide-exporting species in which similar proportions of nitrogen are found distributed among the organic solutes found in xylem sap irrespective of whether the plants are grown on fixed nitrogen or are relying on nitrogen fixation by the nodules for a supply of fixed nitrogen. However, in *Glycine max* (soybean), a ureide-producing species, the effect of supplying fixed nitrogen to the nodulated plant is to alter the pattern of nitrogenous solutes in the xylem fluid from one in which ureide components (produced by nodules) are predominant into one in which amides in particular asparagine produced as a result of the assimilation of combined nitrogen by root tissues, are the predominant transport compounds.

Biosynthesis and properties of ureides

Although there is considerable evidence suggesting that ureide synthesis occurs in legume nodules, the exact location of the pathway in either the bacteroid or nodule cytosol has yet to be confirmed. The ureides allantoin and allantoic acid are synthesised from purines by the degradative pathway illustrated in Fig. 7.2 and these compounds can then be exported from the nodule for use in other parts of the plant. The xylem levels of the ureides allantoin and allantoic acid during plant growth appear to reflect the relative activities of the enzymes uricase and allantoinase responsible for their synthesis in the nodule. High concentrations of allantoin-N have been observed in roots, stems and pods of effectively nodulated soybeans during the time at which pod greening is occurring. Significant amounts of ureides have also been found in leaf tissue but the highest amounts, on an organ basis, are usually found in stem tissue where ureide levels can account for as much as 75 per cent of the soluble nitrogen, as in the case of nodulated soybean plants.

The fraction of the total sap nitrogen composed of ureides can vary very widely depending on the plant species. In *Pisum* about 10 per cent of the total sap nitrogen is made up of ureides while in *Acer* the ureide contribution can be up to 99 per cent of the total sap nitrogen. The fraction of the xylem sap nitrogen made up of ureides remains fairly constant as long as the nitrogen source in the rooting medium remains unchanged.

One advantage to the plant of using ureides as a nitrogen transport molecule has already been mentioned, i.e. the property of these molecules in possessing a low C:N ratio. However, these compounds also have one major disadvantage as transport molecules in that they have a very low solubility compared with amides such as asparagine. Plants which are amide exporters rather than ureide exporters have amide concentrations in the xylem fluid leaving the roots at or near saturation levels. Plants which grow in temperate conditions could not transport the equivalent amount of nitrogen as allantoin as this level of allantoin would be in excess of its solubility. It appears that ureide-exporting plants have overcome this solubility problem by exporting not only allantoin but also allantoic acid, which has a similar solubility to allantoin, thus allowing more nitrogen to be transported in a ureide form than if allantoin alone were used as sole transport compound. In the case of nodulated soybeans, it has been demonstrated that stem exudate contains approximately equal proportions of allantoin and allantoic acid.

The relative insolubility of allantoin has proved beneficial for certain temperate trees, e.g. *Acer* which utilise insoluble allantoin as a nitrogen storage product during winter and then mobilise these reserves in spring when allantoin can be transported to support growth in new shoots. Ureides may also function as transporters of fixed nitrogen from roots to leaves where ureide degradation to NH_3 and CO_2 may occur via pathways shown in Fig. 7.2. The ammonia so produced can be reassimilated into amino acids in leaf tissue. Ureide concentrations are high in young unexpanded leaves but decrease rapidly during leaf expansion before increasing again during leaf maturation.

Allantoin is also used in the production of seed proteins in certain plants. Results indicate that ureides are initially concentrated in pods where there may be some degradation of ureides and assimilation of released 'nitrogen' (NH_3) into other nitrogen-containing compounds, e.g. amides and amino acids. In young pods the ureide levels may account for 50 per cent of the total soluble nitrogen in this tissue. The nitrogenous compounds plus any ureides not degraded can then be transported into the developing seed where ureide degradation continues along with ammonia assimilation into amino acids, purine and pyrimidine bases and subsequently proteins and nucleic acids. The concentration of ureides present in the seed during development reaches a maximum before declining as the seed matures. It has also been observed that the activity of allantoinase, the enzyme involved in the initial step of the pathway leading to NH_3 production from allantoin (Fig. 7.2), is maximal in cowpea fruits at a time when seed storage protein is accumulating rapidly.

Nitrogen metabolism during plant growth

The time at which a plant develops its photosynthetic capability coincides with an alteration in the nitrogen metabolism of the plant. The nitrogenous compounds which have been absorbed, assimilated and metabolised in the roots are exported via the xylem through the stem tissue to the leaves. Knowledge of the nitrogen metabolism of leaf tissue during early growth is sketchy but any nitrogen received by the leaf in excess of its protein synthetic requirements at this time accumulates in the soluble nitrogen pools contained within leaf vacuoles and is not exported via the phloem.

The xylem is the system via which nitrogen is transported into mature leaves, but during transport of nitrogenous materials from root tissue to leaves via the stem there is selective absorption of different amino compounds from xylem fluid by stem tissue. In addition there is also transport of materials from stem tissue to xylem tissue at this time. In the white lupin (*Lupinus albus*), which has been studied in most detail, there is a gradation in the ability of stem tissue to absorb different amino acids. Arginine is very effectively absorbed by stem tissue so that reduced quantities reach the leaves. Asparagine, glutamine, valine and serine are absorbed less efficiently while aspartate and glutamate are absorbed relatively poorly by stem tissue.

The overall effect of a net absorption of nitrogen by stem tissue is a progressive decrease in the nitrogen content of xylem sap further up the stem which would theoretically lead to a reduced supply of nitrogenous materials to leaves higher up the stem. This effect is overcome by an increased transpirational rate in leaves higher up the stem than in leaves in the lower regions of the plant, thus compensating for the more dilute nitrogenous supply in the xylem sap reaching these leaves. Younger unexpanded leaves higher up the stem compensate not by increased transpiration but by an increased reliance on phloem import for their nitrogen demands during initial growth and leaf expansion.

Nitrogenous compounds in leaf tissue are synthesised utilising nitrogen transported into the leaf in the form of amides, amino acids, ureides or nitrate, the types of nitrogen-containing molecules entering the leaf being both characteristic of the species and dependent upon the source of nitrogen available in soil solution.

Plants with low nitrate reductase activity in their roots, e.g. *Xanthium* transport nitrogen mainly in its inorganic form (as

NO_3^-) to the leaf where nitrate reduction occurs prior to its incorporation into amino acids and other nitrogenous compounds utilising carbon skeletons derived from the photosynthetic process in chloroplasts. The products of nitrate assimilation in mature leaves are exported via the phloem for use in other parts of the plant. Young, actively growing regions of the shoot in *Xanthium* are capable of both abstracting nitrate from xylem sap and subsequently reducing it to amino compounds for possible use in shoot growth. Older regions of the stem are less active in nitrate reduction and any nitrate absorbed from the xylem is stored by the stem tissue in this inorganic form.

In the field pea the root system receives a supply of carbohydrate and reductant derived from photosynthetic processes occurring in leaf tissue. These carbohydrates probably supply the carbon skeletons for the synthesis of many of the nitrogenous compounds arising from nitrate reduction and assimilation in root tissue and, in addition, supply the necessary carbon compounds for essential root growth purposes. The reductant is utilised in nitrate reduction by root nitrate reductase. Nitrogenous compounds in the phloem also supply most of the nitrogen requirements for root growth.

The xylem sap leaving the root tissue of field peas contains asparagine, glutamine and ureides as the main nitrogen transport compounds but can also contain some nitrate which has escaped reduction in root tissue. Nitrogenous components being transported to the leaf by the xylem stream can be absorbed by older parts of the stems of field peas and can subsequently be metabolised into a variety of nitrogen-containing compounds, e.g. nucleotides and amino acids by stem tissue. Any nitrate absorbed from the xylem by stem tissue can be reduced by the stem's nitrate reductase prior to subsequent assimilation into organic nitrogen-containing compounds. Nitrogenous compounds may also be released from stem tissue and transported to root and shoot regions.

In leaf tissue, any nitrate entering via the xylem stream can be reduced by the leaf nitrate reductase and used as a nitrogen source in the synthesis of amino acids along with the amide nitrogen entering leaf tissue. The nitrogenous compounds synthesised in mature leaf tissue are exported to growing root and shoot regions primarily as amino acids and amides where, after metabolic interconversions, they can serve as precursors in protein and nucleic acid biosynthesis and for other cellular processes.

In many woody perennials the initial demand for nitrogen for leaf expansion and stem elongation during spring growth is met mainly by the mobilisation and subsequent utilisation of protein reserves stored in the bark over the winter period. This protein is enriched in the amino acid arginine and hydrolysis of the protein causes an initial increase in soluble nitrogen content of the bark (principally amino acids and amides) prior to mobilisation of these compounds to the growing regions of the plant when both protein nitrogen and soluble nitrogen levels in bark decline. The amino acid content of the soluble nitrogen pool does not reflect the amino acid composition of the storage protein indicating that extensive interconversions of amino acids have been undertaken particularly via a channelling of amino nitrogen into asparagine which suggests an important role for this amino acid in the translocation process.

The reserve protein of bark tissue used for the support of spring growth in woody perennials is laid down in the summer and autumn of the previous year. Initially the soluble nitrogenous pool of bark is enriched in the amino acids glutamine and asparagine formed via nitrate reduction and assimilation in the roots prior to transport. The amide nitrogen of these amino acids is then utilised in arginine biosynthesis in the bark prior to incorporation of the amino acid into storage protein. Nitrogenous components for reserve protein biosynthesis may also be supplied by mobilisation of the amino acids resulting from protein hydrolysis in senescing leaf tissue during the autumn. In this way, the nitrogen stored by the woody perennial in the autumn/winter of one year is utilised to support growth during the spring of the following year.

Nitrogen mobilisation during senescence

Senescence in the whole plant arises from a series of programmed physiological changes which are thought to be controlled by nuclear genes. A leaf undergoes a series of developmental changes during its lifetime prior to senescence. Initially the leaf is a predominantly heterotrophic organ which then matures to become a net exporter of photosynthate. After reaching maturity, the leaf exhibits a reduced capacity to act as a source of photosynthetically fixed carbon and enters a senescent phase when it becomes a major source of nitrogen and minerals as its reserves are depleted and transferred to other parts of the plant. At the termination of the process of senescence the leaf is abscinded.

The organelles of a senescing leaf do not all degenerate at the same time but exhibit an ordered sequential breakdown. The first signs of organelle degeneration are apparent in free ribosomes and chloroplasts and then at a later stage in mitochondria. The final stages of senescence are characterised by the disruption of the nucleus and plasmalemma. These ordered ultrastuctural changes are also accompanied by changes in both leaf nitrogen content and in the composition of nitrogenous components of leaf tissue. A controlled proteolysis of leaf protein precedes any visible signs of senescence in the leaf and is accompanied by chlorophyll loss and a decline in the total leaf RNA content and an associated decrease in the rate of RNA synthesis in the leaf during this period.

Amino acid nitrogen arising from protein degradation in leaf tissue is channelled into the formation of the amino acids asparagine and glutamine prior to export from the leaf via the phloem. In addition there appears to be extensive interconversion of amino acids in senescing leaves during this period with some of the carbon skeletons of the amino acids being utilised as respiratory substrates in the leaf. The overall effect of these degradative processes is to decrease the nucleic acid, protein and soluble nitrogen content of senescent leaf tissue as the soluble nitrogenous transport components, in particular asparagine and glutamine, are exported from the leaf tissue via the phloem to other parts of the plant. This mobilisation of nitrogen causes an alteration in the composition of the phloem sap from one in which carbohydrates predominate into one in which amino acids are the predominant components.

The fate of these nitrogenous components from senescent leaves depends upon the type of plant under consideration. The supply of nitrogenous compounds to the developing fruits and seeds of annual plants relies very heavily on the mobilisation of leaf-bound nitrogen to the phloem during leaf senescence. In deciduous woody species the nitrogen mobilised from leaf tissues during senescence is transported by the phloem to the trunk where it is used in the synthesis of food reserves in the bark. In herbaceous plants the mobilised leaf nitrogen is utilised for the production of storage reserves in stems and storage organs, these reserves often including significant amounts of nitrogen in soluble organic form.

The principal nitrogenous transport products in plants may not always be the form in which the plant stores nitrogen. Asparagine is the principal form by which nitrogen from insect digestion is translocated in *Drosera erythrorhiza*, but the tubers of this plant utilise arginine as the principal nitrogenous storage form.

Nitrogen metabolism during seed development

Developing fruits and seeds rely heavily on the nitrogen supplied in organic forms in the phloem translocate for their nitrogen nutrition. Studies on the flow of C, N and water in xylem and phloem during fruit development in *Lupinus albus* have suggested that 98 per cent of the carbon, 89 per cent of the nitrogen and 40 per cent of the water entering the fruit are supplied by the phloem, the remaining nitrogen and water being supplied by the xylem. The xylem and the fruit's own photosynthetic activity supply the remaining carbon. Three compounds in particular play a major role in carbon and nitrogen supply to the fruit in *Lupinus albus*. Sucrose comprises 90 per cent of the phloem's carbon supply to the fruit while the amino acids asparagine and glutamine comprise 75–85 per cent of the nitrogen supply in the phloem and xylem. When these amino acids reach the developing fruits they undergo extensive metabolic interconversions, particularly through the channelling of their nitrogen into the amino acid arginine which, although virtually absent in the phloem, is found as a major amino acid component of the seed's storage proteins.

Studies on the metabolism of asparagine in developing seeds of *Lupinus albus* have shown that while the seed receives 55–60 per cent of its nitrogen as asparagine via the phloem, the seed protein formed from the nitrogen supply contains only 7–10 per cent of asparagine residues. An extensive metabolism of asparagine has taken place through which at least 60 per cent of the amide nitrogen has been directed into the synthesis of many different amino acids especially arginine at a time of maximal synthesis of storage protein. Very high levels of asparaginase have been detected in developing seeds of *Pisum sativum* and the ammonia released from asparagine by this enzyme could then be reassimilated by glutamine synthetase and GOGAT (see p. 27) prior to incorporation into other amino acids which are synthesised *de novo* during seed formation. This extensive redistribution of nitrogen in the developing seed is necessary because the composition of nitrogenous compounds delivered to the developing seed in the phloem and xylem is vastly different to that required by the seed for synthesis of seed proteins and nucleic acids. These interconversions are summarised in Fig. 7.3.

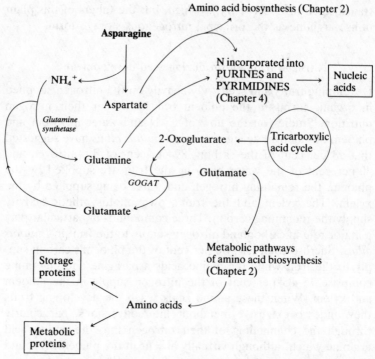

Fig. 7.3 Interconversions of asparagine-N in developing seeds

The main source of nitrogen for the developing fruits and seeds of annual plants in which leaf senescence accompanies fruit and seed development is the supply of mobilised leaf-N in the phloem. However, leaf senescence does not always accompany fruit and seed development as is the case with many perennials where the nitrogen for amino acid biosynthesis in the developing fruit seed is derived from the nitrogenous components translocated in the phloem from leaves which are actively photosynthesising. Other exceptions include some spring flowering trees, e.g. elm, where leaf development does not occur until after seed production. These plants mobilise their nitrogenous reserves components, e.g. those in the bark such as arginine, to supply the nitrogenous requirements of their developing fruits and seeds.

Nitrogenous reserves in seeds

The chemical composition of a mature seed is determined mainly by the nature of its principal food reserves which, depending on the seed, may be contained mainly in either the endosperm or the cotyledons of the embryo.

The protein reserves found in seeds are characteristic of a particular species, in dicotyledons they are principally globulins, e.g. legumin and vicilin found in peas and beans, while in cereals the main storage proteins are essentially prolamines (proline and glutamine rich; lysine poor), e.g. zein from maize, and glutelins which are complex proteins of high molecular weight characteristic of their source of origin. In addition to reserve proteins, seeds also possess many enzyme proteins responsible for the operation of the many general and specific synthetic and degradative processes operative in the developing fruit and seed. Reserve proteins are not always distributed uniformly throughout the cell but are often stored in subcellular structures called protein bodies, e.g. legumin and vicilin are found in protein bodies in the cotyledons of legumes.

The nucleic acid content of seeds varies widely depending upon species. When the reserve tissue is cotyledonary as in the case of legumes, there is a relatively high level of nucleic acid, particularly in the form of ribosomal RNA in ribosome particles in this tissue. This situation contrasts with that of cereals where reserve materials are stored in the endosperm which contains very little nucleic acid although the embryos of cereal seeds, which in the case of wheat seed comprise on average only 2–3 per cent by weight of the grain, do contain high levels of both DNA and RNA.

Biosynthesis of seed proteins

The mobilisation of the soluble nitrogen components of the developing seed into the protein and nucleic acid components found in the mature seed follows a pattern which is characteristic of that particular plant. Mature seeds contain only a small proportion of their nitrogen as soluble amino acids whereas developing seeds contain an amino acid pool which is constantly changing in both size and composition during maturation of the seed. Amino acid interconversions are performed by the many enzymes of amino acid metabolism discussed in previous chapters, e.g. asparaginase and the many transaminases which are present at high levels of activity in developing seeds. In general the level of amino acids in the developing seed increases coincidentally with the period at which protein synthesis reaches a maximum and decreases as the rate of protein synthesis decreases.

The deposition of nitrogenous reserves in legumes and cereal seed follows a different pattern. In peas, there is an initial development of the pod prior to seed enlargement and the amino acids which accumulate in the pod along with those which accumulate initially in the endosperm (the endosperm components being mobilised to support the growth of the developing embryo in peas) contributes a significant proportion, possibly up to one-fifth, of the amino acid nitrogen used in protein synthesis in the developing embryo. The initial proteins which accumulate in developing pea embryos are albumins while the storage proteins – vicilin and legumin – are synthesised in the later stages of seed development.

The synthesis of RNA in developing pea seeds continues after cell division has ceased and remains constant during the period of dehydration of the seed. Surprisingly, DNA synthesis in peas and other legumes continues after cell division has ceased in the developing embryo with the result that pea seeds may contain cells with up to 64C in comparison to the expected 2C diploid levels of DNA. The DNA levels of the cotyledon cells remain constant during the period of dehydration of the seed.

In maize, the synthesis and deposition of protein during seed development occurs in two phases. The protein accumulates in the endosperm rather than in the embryo where the accumulation of nitrogenous reserves is very small. Nucleic acids also accumulate in endosperm tissue during the initial phase of cereal seed development but in contrast to the situation in developing legume seeds, the DNA content of cereal seeds reaches a maximum when cell division ceases while the RNA content of the endosperm actually declines during the second phase of protein deposition in maize endosperm. Mature cereal seeds have a relatively low level of nucleic acids in the endosperm with the nucleic acids concentrated mainly in the embryo portion of the grain.

It has been suggested that in many seeds, where a large proportion of the ribosome population has been observed to be membrane bound during seed development, reserve proteins are synthesised on these membrane-bound ribosomes prior to transport across the membrane into the lumen of the endoplasmic reticulum. These proteins would be subsequently transported, by a means not yet determined, into the protein bodies of the seed. The proposed mechanism for ribosome attachment to membranes and the transport of storage proteins across membranes and into the lumen of the endoplasmic reticulum is via the 'signal hypothesis' (see p. 178) which also accounts for the synthesis of secreted cellular proteins.

The mature seed is now in a state whereby, once any dormancy has been broken and suitable conditions for germination prevail, the whole cycle of growth, development and senescence with their accompanying phases of nitrogen fixation, mobilisation and inter-conversions linked to the multitude of metabolic reactions involving the nitrogen atom in plant tissues can again proceed.

Suggestions for further reading

Anderson, J. D. (1979) Purine nucleotide metabolism of germinating soybean embryonic axis, *Plant Physiol.*, **63**, 100–4.

Bewley, J. D. & Black, M. (1978) Physiology and Biochemistry of Seeds in Relation to Germination. Vols. 1 and 2. Springer-Verlag: Berlin, Heidelberg, New York.

Bray, C. M. (1979) Nucleic acid and protein synthesis in the embryo of germinating cereals, in D. L. Laidman, and R. G. Wyn-Jones (eds) *Recent Advances in the Biochemistry of Cereals*. Academic Press, 1979.

Miflin, B. J. & Lea, P. J. (1977) Amino acid metabolism, *Ann. Rev. Plant Physiol.*, **28**, 299–329.

Pate, J. S. (1980) Transport and partitioning of nitrogenous solutes, *Ann. Rev. Plant Physiol.*, **31**, 313–40.

Rawsthorne, S., Minchin, F. R., Summerfield, R. J., Cookson, C. & Coombs, J. (1980) Carbon and nitrogen metabolism in legume nodules, *Phytochem.*, **19**, 341–55.

Rodaway, S. & Marcus, A. (1979) Germination of soybean embryonic axes. Nucleotide sugar metabolism and initiation of growth, *Plant Physiol.*, **64**, 975–81.

Sprent, J.I. (1980) Root nodule anatomy, type of export product and evolutionary origin in some leguminosae, *Plant Cell and Environment*, **3**, 35–43.

Thomas, R. J. & Schrader, L. E. (1981) Ureide metabolism in higher plants, *Phytochem.* **20**, 361–71.

Index